Springer Series in Advanced Manufacturing

Series Editor

Duc Truong Pham, University of Birmingham, Birmingham, UK

The **Springer Series in Advanced Manufacturing** includes advanced textbooks, research monographs, edited works and conference proceedings covering all major subjects in the field of advanced manufacturing.
The following is a non-exclusive list of subjects relevant to the series:

1. Manufacturing processes and operations (material processing; assembly; test and inspection; packaging and shipping).
2. Manufacturing product and process design (product design; product data management; product development; manufacturing system planning).
3. Enterprise management (product life cycle management; production planning and control; quality management).

Emphasis will be placed on novel material of topical interest (for example, books on nanomanufacturing) as well as new treatments of more traditional areas.

As advanced manufacturing usually involves extensive use of information and communication technology (ICT), books dealing with advanced ICT tools for advanced manufacturing are also of interest to the Series.

Springer and Professor Pham welcome book ideas from authors. Potential authors who wish to submit a book proposal should contact Anthony Doyle, Executive Editor, Springer, e-mail: anthony.doyle@springer.com.

More information about this series at http://www.springer.com/series/7113

Kapil Gupta · Munish Kumar Gupta
Editors

Optimization of Manufacturing Processes

Springer

Editors
Kapil Gupta
Department of Mechanical and Industrial Engineering Technology
University of Johannesburg
Doornfontein, Johannesburg, South Africa

Munish Kumar Gupta
Assistant Professor, University Center for Research and Development
Chandigarh University, Gharuan
Mohali, India

ISSN 1860-5168 ISSN 2196-1735 (electronic)
Springer Series in Advanced Manufacturing
ISBN 978-3-030-19637-0 ISBN 978-3-030-19638-7 (eBook)
https://doi.org/10.1007/978-3-030-19638-7

This Springer imprint is published by the registered company Springer Nature Switzerland AG
The registered company address is: Gewerbestrasse 11, 6330 Cham, Switzerland

Preface

In this era of the Fourth Industrial Revolution, intelligent manufacturing is a global trend to simultaneously optimize quality, productivity, and sustainability. Optimization of manufacturing processes and systems is possible with the available Industry 4.0 tools and techniques. This book provides a detailed understanding on optimization of various manufacturing processes and systems using some of the important statistical and evolutionary (soft computing or can be called as Industry 4.0 based) techniques. It covers sufficient theoretical details, salient features, implementation steps, effectiveness and outcomes of statistical, multi-criteria decision-making and evolutionary techniques for single and multi-objective optimization to improve quality, productivity, and sustainability in manufacturing.

This book consists of nine chapters on optimization of manufacturing processes. Chapter "Modelling and Optimization of Alpha-set Sand Moulding System Using Statistical Design of Experiments and Evolutionary Algorithms" sheds light on modelling and optimization of sand moulding system by GA-, PSO-, and TLBO-type evolutionary techniques. Chapter "Optimization of Electric Discharge Machining Based Processes" provides comprehensive information on optimization of electric discharge machining-based processes via theory, the literature review, and case studies. Chapter "Optimization of Accuracy and Surface Finish of Drilled Holes in 350 Mild Steel" details a statistical optimization of hole quality characteristics in drilling of mild steel. A case study on response surface methodology-based modelling and desirability optimization of laser additive manufacturing of titanium is discussed in Chapter "Modelling and Optimization of Laser Additive Manufacturing Process of Ti Alloy Composite". Chapter "Prediction and Optimization of Tensile Strength in FDM based 3D Printing Using ANFIS" discusses the optimization of mechanical properties of 3D printed parts using ANFIS. Chapter "Optimization of Abrasive Water Jet Machining for Green Composites using Multi-variant Hybrid Techniques" focuses on multi-objective optimization of abrasive water jet machining with the help of MOORA, GA, TOPSIS, and DEAR methods. Machining condition optimization while turning titanium using integrated fuzzy MOORA method is given in Chapter "An Integrated Fuzzy-MOORA Method for the Selection of Optimal Parametric

Combination in Turing of Commercially Pure Titanium". Chapter "Application of Multi-objective Genetic Algorithm (MOGA) Optimization in Machining Processes" provides a review of the literature on implementation of genetic algorithm (GA) for quality optimization of different machining processes. Chapter "Optimization in Manufacturing Systems Using Evolutionary Techniques" focuses on optimization of manufacturing systems by GA and particle swarm optimization (PSO) techniques.

We sincerely acknowledge Springer for this opportunity and their professional support. Finally, we would like to thank all the chapter contributors for their availability and valuable contributions.

Johannesburg, South Africa
March 2019

Kapil Gupta
Munish Kumar Gupta

Contents

Modelling and Optimization of Alpha-set Sand Moulding System Using Statistical Design of Experiments and Evolutionary Algorithms

G. C. Manjunath Patel, Ganesh R. Chate and Mahesh B. Parappagoudar

Abstract The traditional trial-and error method applied to derive empirical relation and optimize the process is time consuming and results in reduced productivity, high rejection and cost. Hence, current research in foundries focussed towards development of statistical modelling and optimization tools. The present research work is focused on modelling and optimization of Alpha-set moulding sand system. The variables such as percent of resin and hardener, and curing time will influence the sand mould properties, namely, compression strength, permeability, mould hardness, gas evolution and collapsibility. Experimental data is collected as per CCD design matrix and non-linear models have been developed for all responses. The behaviour of all responses is studied by utilizing surface plots. The statistical adequacy of all models is tested with help of ANOVA. All responses are tested for their prediction capacity with the help of test cases. The predictive non-linear models, developed for the process resulted in average deviation of less than 5%. The optimization (GA, PSO, DFA and TLBO) tools are applied to optimize the process for conflicting requirements in sand mould properties. Six case studies with different combination of weight fractions assigned to sand mould properties are considered. The optimum solution correspond to highest composite desirability value is selected. TLBO outperformed other optimization tools (i.e. GA, PSO, and DFA) while determining the highest desirability value and resulted in optimized sand mould properties. Experiments are conducted for the optimized and normal (i.e. lowest desirability) sand mould conditions. Castings are prepared by pouring molten LM20 alloy to the prepared moulds. The casting obtained for the optimized sand mould condition resulted in a better casting quality.

G. C. Manjunath Patel (✉)
Department of Mechanical Engineering, PES Institute of Technology and Management, Shivamogga 577204, Karnataka, India
e-mail: manju09mpm05@gmail.com

G. R. Chate
Department of Industrial and Production Engineering, K.L.S. Gogte Institute of Technology, Belgaum, India

M. B. Parappagoudar
Department of Mechanical Engineering, Padre Conceicao College of Engineering, Verna, Goa, India

K. Gupta and M. K. Gupta (eds.), *Optimization of Manufacturing Processes*, Springer Series in Advanced Manufacturing, https://doi.org/10.1007/978-3-030-19638-7_1

Keywords Casting · Moulding · Design of experiments · Genetic algorithm · Teaching learning-based optimization · Particle swarm optimization

1 Introduction

In sand castings, parts and components are produced by pouring molten metal into the sand mould. The quality of casting is largely influenced by the moulding sand properties. Hence, attaining good moulding sand properties is of industrial relevance. Holtzer et al. reported that, approximately 103 million tonnes of metal was accounted for the production of cast parts throughout the year across the globe [1]. 80% production of cast parts were produced in sand moulds with either bentonite or organic resin binder. Resin bonded sand mould (that is, chemical binder) system offered better sand mould and casting properties as compared to green sand moulds (that is, clay binder or bentonite) [2]. Thoroughly mixed silica sand with chemicals (i.e. binder) will help to harden the mould using catalytic reaction [3]. Chemically mixed silica sand has the ability to prepare moulds with intricate shape and precise dimension at an ambient temperature [4]. However, chemical mixed sand moulds emit harmful toxic gases (i.e. chemical compounds) during the foundry processes, which causes environmental pollution and serious human health hazards. The emissivity of harmful compounds (benzene, toluene, ethylbenzene and xylene (BTEX) and polycyclic aromatic hydrocarbons (PAHs) group) to environment from alphaset resin is 2–5 times lesser than the furan resin sand moulds [5]. Further, smokeless, minimal erosion, better hot strength, good finishing and better collapsibility are the distinguished characteristics of alphaset binder [5, 6].

The molten metal poured into the mould cavity had released the undesirable gasses and resulted many surface defects in the casting produced [7–9]. Alphaset binder sand moulds offer excellent surface quality on ferrous castings, due to the absence of phosphorous and sulphur [6]. Further, absence of nitrogen restricts the formation of pinholes in the casting part [6]. Alphaset binder is found to be eco-friendly and keeps the foundry with a healthy working environment. Hence, working on Alphaset bonded sand moulding system to produce good quality castings is of industrial relevance.

The detailed analysis of moulding sand system with associated sand mould properties will provide good insight of a process and on casting quality. Chemical mix sodium silicate sand moulds will have low quantity of gas evolution (GE) as compared to the green sand moulding [7]. Sodium silicate binder is relatively cheap, but is limited to high residual stress, poor shake out property (that is, collapsibility) and difficult to sand reclamation [10]. The moulds with a poor collapsibility (CP) will result in casting defect, namely hot tear [11]. Casting dimensional accuracy and surface finish are primarily dependent on the mould compression strength and hardness. Correlation among sand mould properties were studied by researchers in the recent past. Strong third order non-linear regression relationship exists between the compression strength (CS) and mould hardness (MH) in sand moulds [12]. The

CS of sand moulds increased with the increase in MH, this was due to the existence of strong dependency relationship among themselves. Lower compression strength will yield rough casting surface, shrinkage porosity, sand erosion and dimensional inaccuracy etc. However, high compression strength moulds do not allow the generated gas to escape from mould (i.e. permeability, P). Casting defects (i.e. blowhole, misrun and porosity) in sand moulds might occur as a result of insufficient space for the trapped or generated gases inside the sand mould [13]. The casting quality in sand moulding process is affected largely by the moulding sand properties. The inappropriate combination of moulding sand properties will result in casting defects such as, blow holes, rat tails, misruns, dimensional in accuracies, rough surface, porosity, segregations and so on [14]. These defects can be minimized by selection of optimal levels of moulding sand variables (that is, grain fineness number, degree of ramming, percent of resin and hardener, curing time and so on). Observations made from the above literature, shows that the casting quality is dependent primarily on sand mould properties (GE, CS, MH, CP, and P). Further, studying the appropriate method to control the moulding sand properties is of significant scope for the researchers.

Research work on moulding sand system in the past few decades was more focused on classical engineering experimental (that is, varying one parameter at once after fixed the rest at middle values), analytical and numerical approaches. Numerical methods were applied to predict the gas evolution when the molten metal was poured into the furan bonded moulding sand system [15]. The chemically mixed sand moulds contain resin and hardener, however their impact in furan sand moulds on gas evolution was neglected and was limited to establish the input-output relationships. Classical experiments were conducted to study the influence of different quantity of furan resin and hardener on moulding strength, gas evolution, surface quality and casting microstructure [16–18]. However, interaction effects of furan resin and hardener quantity were not considered and no predictive equations were developed in their study. The optimized binder composition had yielded good casting surface features with dimensional accuracy and better mould collapsibility [19]. However, the effect of curing time was not considered during their experimental investigations. The effect of sodium silicate and bentonite binders on gas evolution was studied by conducting classical experiments [7]. The influence of size (that is, coarse or fine) of the sand particles and their impact on moulding sand permeability and casting surface finish studied [11, 20]. The coarse sand resulted in a high permeability with rough casting surface, whereas, smooth uniform casting surface was obtained with fine sand particles but resulted in low permeability. Gas porosity in castings occurred due to the generated pressure inside the mould as a result of low permeability [13]. High percent of resin was resulted in better mechanical properties in sand moulds, whereas it was difficult to extract the cores from the solidified metal cast [21]. Further, higher resin content resulted in evolving huge amount of gas due to resin decomposition during casting solidification. The evolved gases resulted in defects in the casting part [21]. The above literature confirmed that, the moulding sand variables (proportion of resin and hardener, curing time, grain fineness number and so on) will have large influence on sand mould properties and in-turn casting quality. The conflicting requirements (higher compression strength in moulds offer low permeability and vice versa) in sand mould are found to be complex (that is, non-linear). It is difficult to establish good control of a process with analytical, numerical and classical experiment approaches. To attain better casting quality through setting the sand mould properties at optimal level by utilizing appropriate method (that is, modelling and analysis) is the concern for foundry industries.

Early in 1964, the statistical methods were first applied to conduct foundry experiments and analysis [22]. Design of experiments (DOE) is an effective statistical tool used to study, analyse and establish the input-output relationships. DOE methods estimate both individual and combined factor effects by conducting minimum experimental trials. Taguchi method was applied to model and optimize the casting defects in green sand moulding process [23]. Computer aided simulation was performed to obtain the casting data (defects) for Taguchi parametric design. Taguchi method might fail to establish full quadratic (that is, linear, square and interaction) factor effects due to limitation in orthogonal array. DOE was applied to conduct analysis of moulding sand variables (proportion of resin, and hardener, and temperature) on moulding strength and sand inclusion defects in furan sand moulding system [24]. DOE and response surface methodology (RSM) were applied to get the full quadratic factor effects and develop predictive models for green sand moulding [25], phenol formaldehyde sand mould [26], sodium silicate, CO_2 gas hardened [27], and furan sand moulding processes [28]. Further, the models predicted the sand moulding properties with a better accuracy for test cases. It was observed from the above literature that, the design of experiments and response surface methodology is an ideal tool to study many variables, which are complex and highly non-linear. Further, the combined tools can be used to establish precise mathematical equation representing outputs as a function of inputs. Moreover, the derived empirical relationship can be used to determine optimal points for a process. Important to note that, not much of the research efforts were made in the recent past on modelling, analysis and optimization of Alphaset sand moulding system.

The optimization task is generally carried out to determine the best results subjected to various resource (that is, input or design variable) constraints. Conventional optimization and nonconventional optimization are the two broad classifications of optimization techniques, distinguished based on search (that is, operating) mechanism employed to yield best results [29]. Conventional optimization algorithms are deterministic algorithms (such as, dynamic programming, non-linear programming, quadratic programming, geometric programming etc.) work with specific transition rules for moving solutions from one to another space during the optimal search [30]. For multi-modal optimization problems, the conventional optimization methods fail to locate the global optimal solutions. This is due to the difficulty in handling many variables with complex non-linear characteristics. The speed of convergence to locate the optimal solutions with conventional optimization tools is slow. Nonconventional optimization tools (that is, population based search methods include evolutionary and swarm intelligence algorithms) overcome the difficulties of conventional optimization tools by attaining the global solutions and rapid convergence that yield better results. Nonconventional optimization tools use heuristic search methods with definite set of probabilistic transition rules to get better solutions. The difference in the performance of evolutionary and swarm intelligence algorithms can be found for multi-modal and multidimensional optimization problems. This occur due to the different search mechanisms and different combination of employed rules to move population and associated solutions towards optimal. The population based algorithms [genetic algorithm (GA), particle swarm optimization (PSO), and teaching learning based optimization (TLBO)] are cost effective optimization tools in determining near optimal solutions through their heuristic search mechanism. Evolutionary GA, needs to set mutation rate and crossover parameters, and Swarm intelligence based PSO, needs to set inertia weight, social and cognitive parameters, at optimal level to yield best results [31]. Improper choice of genetic and swarm optimization parameters will affect both the computational efficiency and optimality of solutions [30]. Teaching learning based optimization (TLBO) do not require specific tuning parameters, thereby the probability to hit the global solutions are more [32]. GA and PSO were applied to optimize the green sand moulding [14] and squeeze casting process [33, 34] for better casting quality. PSO and GA had produced approximately similar results while locating global solution, and the computational effort and time was less for PSO. TLBO outperformed GA, PSO, and Taguchi optimization tools while performing optimization for different casting (that is, squeeze casting, die casting and continuous casting) and machining (wire electric discharge machining, abrasive jet machining and ultrasonic machining) processes [35, 36]. GA, PSO and TLBO tools can be applied to optimize the conflicting requirements (that is, maximize: CS, MH and P, and minimize: CP and GE) in sand mould properties. Further, use of optimization tools will minimize the requirements of practical experiments and analytical tools which are always costlier, tedious and time-consuming.

In the present work, the modelling of eco-friendly alphaset bonded sand mould system is conducted to understand the effect of sand moulding variables and moulding sand properties and to establish accurate relation between them. Statistical analysis will help the foundry personnel and researchers to study the full quadratic factors effects (linear, square and interaction) on sand mould properties. The statistical and 3D surface plot analysis will provide detailed insight of the physics of a process (i.e. process mechanics and dynamics). Further, sand mould properties (CP, CS, P, GE and MH) are expressed as a mathematical non-linear function of input variables. These predictive equations will help the foundry man to know the values of sand mould properties for the known set of moulding sand variables (that is, percent of resin, percent of hardener, and setting time). The conflicting requirements in moulding sand properties (minimize: GE and CP and maximize: CS, MH and P) are optimized by applying non-conventional optimization methods (that is, GA, PSO, and TLBO). Motivated by this, systematic study of modelling and optimization of alphaset bonded sand mould system would help the foundry personnel to obtain good quality castings, without much efforts, time and prior detailed knowledge of the process.

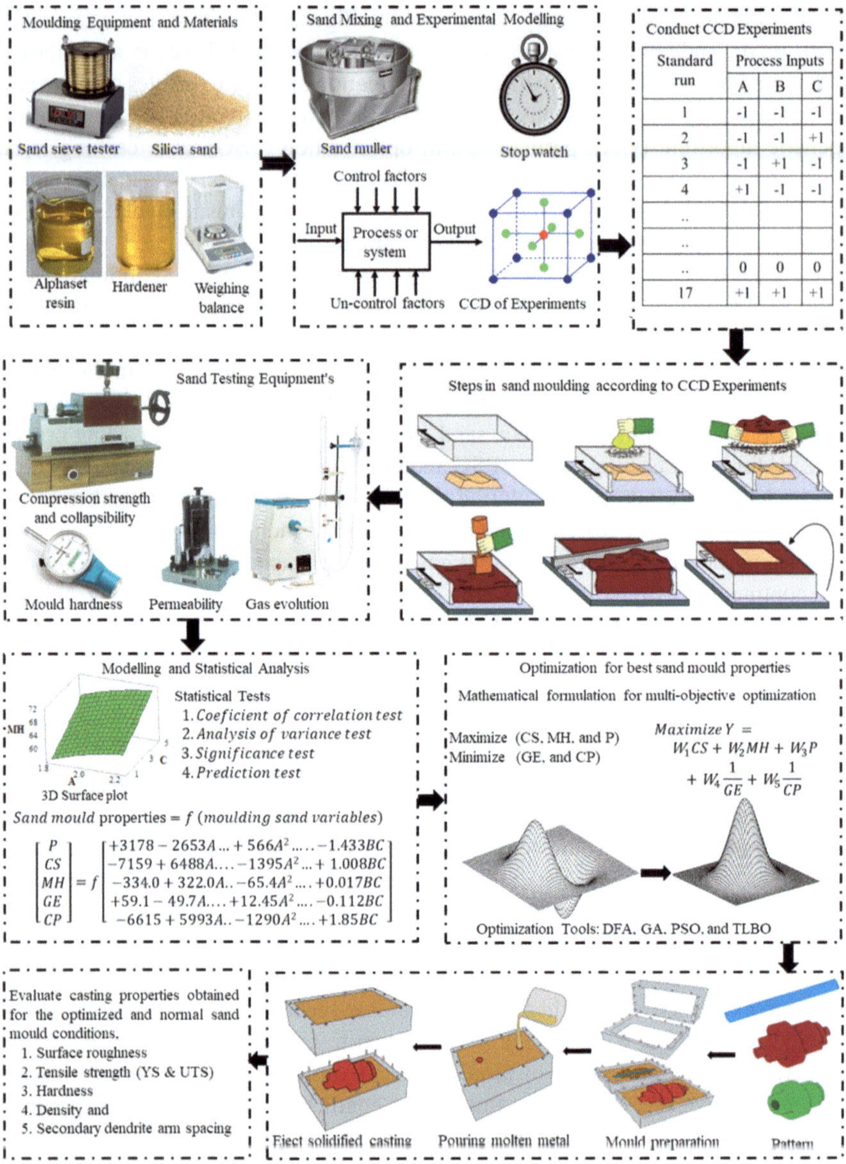

Fig. 1 Sequence of tasks performed during experimentation, modelling and optimization for better sand mould and casting properties

2 Experimentation, Modelling and Optimization

The experiments have been conducted in Alphaset sand moulding process. Further, the experimental data is used to develop non-linear models and optimize the process parameters. Figure 1 shows the sequence of various tasks with experimental setup during experimentation, modelling, and optimization of Alphaset sand moulding.

Step 1: The moulding materials (i.e. alphaset resin, hardener, and silica sand) are collected for experimentation. Sieve analysis test is conducted to determine the grain fineness number (GFN) as per American Foundry Society (AFS) standard. The required quantity of resin and hardener is measured with the help of digital weighing balance. The moulding materials and associated parameters used for the experimentation are selected based on trial experiments, consulting industrial expert's opinion, and available literature [12, 16–19, 25–28] (refer Table 1).
Step 2: Experiments are conducted with different set of sand mould variables as per Central Composite Design (CCD) matrix. The specimens (height of 5 cm and 3 cm in diameter) are prepared in accordance with American Foundry Society standard. The test specimens prepared as per CCD are used to determine the sand mould properties (that is, CS, P, MH, CP and GE).
Step 3: Moulding sand properties are expressed in terms of moulding sand variables by non-linear mathematical equations (regression models). Statistical tests (that is, coefficient of determination, significance test, analysis of variance and prediction tests) are carried out to determine statistical adequacy and to understand the behaviour of variables on mould properties.
Step 4: The derived empirical relationships for all sand mould properties are treated as an objective function for process optimization. Weight based method is employed to convert the conflicting objective function (maximize: CS, MH and P, and minimize: GE and CP) to a single objective function for maximization. GA, PSO, and TLBO algorithms are applied to optimize the sand mould properties in Alphaset sand moulding process. Further, the casting quality is evaluated for the different sand moulding conditions (that is, optimized and normal).

Table 1 Moulding materials and associated parameters

Parameters	Value
Grain fineness number or AFS number	55
Alphaset resin	1.8–2.2%
Ester cured	0.2–0.4%
Weighing balance accuracy	0.1 mg
Degree of ramming	3
Curing time	60–120 s
Mulling time	180 s

Table 2 Sand mould variables and associated operating levels

Input variables	Units	Un-coded levels		
		Low	Medium	High
[a]Resin	%	1.8	2.0	2.2
[a]Hardener	%	0.2	0.3	0.4
Curing time	min	60	90	120

[a]wt% of sand

3 Data Collection

The silica sand with 55 GFN is mixed with binder and catalyst for 3 min in sand muller and the test specimen are prepared by using this moulding sand mixture. The operating levels of moulding sand (input) variables used for conducting the experiments is presented in Table 2.

The following tests are conducted on the prepared test specimen to measure the sand mould properties. The height of the sand mould specimens are measured using standard height gauge and kept within the range of 5–5.1 cm. The permeability measurements are conducted using permeability meter. The compression strength and collapsibility (that is, retained strength) in the sand moulds are measured with the help of universal strength testing unit. The samples are kept in a muffle furnace maintained at 650 °C in for period of about 2 min and the collapsibility (that is, strength retained after heating) is determined by using universal strength testing machine. The harmful toxic compounds are emitted when the resin comes in contact with the molten metal. This will pollute the environment and cause serious threat to human health. 1 g of thoroughly mixed silica sand with alphaset binder and hardener, is taken in the ceramic boat and placed in the heated tube maintained at a temperature of 850 °C. The evolved gases as a result of burnt resin and hardener is measured by the displacement of water level in burette. The unit corresponding to gas evolution is ml/gm (Table 3).

4 Analysis, Modeling and Optimization

This section describes the modelling and optimization of Alphaset sand moulding process. The input-output data collected through experiment is used to develop surface plots for responses namely, GCS, CP, GE, MH, and P. The surface plots are the powerful graphical tool depicting the influence of linear and non-linear relation of the responses with input parameters. Statistical analysis is conducted to know the significance of full quadratic factor effects (that is, linear, square and interaction) of moulding sand variables on sand mould properties is tested by analysis of variance (ANOVA). Minitab (version 17) platform software is utilized for the said purpose. The prediction tests are also conducted to check the performance and practical utility

Table 3 CCD based experimental matrices for alphaset sand moulding

Exp. No.	Input variables			Sand mould properties				
	A	B	C	CS, KPa	CP, KPa	GE, ml/gm	MH	P
1	2.0	0.3	090	371.2	270.70	7.60	75.1	118.5
2	2.2	0.4	120	388.9	285.20	9.28	75.6	106.6
3	2.0	0.3	120	378.6	280.20	7.75	76.8	125.8
4	2.0	0.3	060	350.7	268.20	7.34	72.8	134.6
5	1.8	0.4	060	252.8	181.70	6.63	67.1	159.5
6	2.2	0.4	060	390.7	290.30	9.01	73.1	108.9
7	2.0	0.2	090	351.8	257.50	8.70	71.8	131.9
8	1.8	0.2	060	168.6	107.80	8.54	60.7	203.5
9	2.2	0.2	060	368.7	280.50	8.86	74.2	109.3
10	2.0	0.4	090	421.6	328.20	8.49	77.7	101.4
11	2.0	0.3	090	372.8	274.50	8.27	75.2	114.2
12	2.2	0.2	120	360.6	272.04	8.11	74.7	138.4
13	1.8	0.4	120	352.8	264.60	8.52	73.9	131.6
14	2.0	0.3	090	370.7	288.70	8.43	75.5	116.7
15	2.2	0.3	090	372.2	281.20	8.97	75.6	121.3
16	1.8	0.2	120	250.7	149.60	8.78	69.1	178.6
17	1.8	0.3	090	268.8	182.60	8.32	70.5	171.2

of the developed models. Castings are prepared by pouring the molten metal into the sand mould prepared with normal and optimized conditions and evaluated the casting quality.

4.1 Analysis and Modeling

4.1.1 Compression Strength

The compression strength, expressed as a non-linear mathematical function of moulding sand variables is as follows,

$$\begin{aligned} \text{CS} = & -7159 + 6488\text{A} + 1291\text{B} + 10.69\text{C} - 1395\text{A}^2 + 1040\text{B}^2 - 0.01294\text{C}^2 \\ & - 850\text{AB} - 4.0\text{AC} + 1.008\text{BC} \end{aligned} \tag{1}$$

Significance tests are carried out for the developed non-linear model to check the contributions of full quadratic (linear, square and interaction) effects of the factors on compression strength. The significance tests are conducted for the preset (P-value ≤ 0.05) confidence level of 95%. The significance test results obtained

Table 4 ANOVA test results for sand mould properties

Response		Compression strength				Collapsibility			
Source	DF	Adj. SS	Adj. MS	F	P	Adj. SS	Adj. MS	F	P
Model	9	68,492.5	07610.3	161.96	0.000	54,232.1	06025.8	062.71	0.000
Linear	3	47,896.0	15965.3	339.77	0.000	36,847.0	12282.3	127.81	0.000
Square	3	13,603.3	04534.4	096.50	0.000	11,306.6	03768.9	039.22	0.000
Interaction	3	06993.2	02331.1	049.61	0.000	06078.6	02026.2	021.09	0.001
Error	7	00328.9	0047.0			00672.7	00096.1		
Lack of fit	5	00326.5	0065.3	054.27	0.018	00492.6	00098.5	001.09	0.541
Pure error	2	00002.4	0001.2			00180.0	00090.0		
Total	16	68,821.4				54,904.8			
Response		Mould hardness				Gas evolution			
Model	9	272.419	30.269	29.55	0.000	6.85901	0.76211	12.04	0.002
Linear	3	179.606	59.069	58.45	0.000	1.72008	0.57336	09.06	0.008
Square	3	057.943	19.314	18.86	0.001	1.87379	0.62460	09.87	0.007
Interaction	3	034.870	11.623	11.35	0.004	3.26514	1.08838	17.19	0.001
Error	7	007.170	01.024			0.44309	0.06330		
Lack of fit	5	007.083	01.417	32.69	0.030	0.05529	0.01106	00.06	0.995
Pure error	2	000.087	00.043			0.38780	0.19390		
Total	16	279.589				7.30209			
Response		Permeability							
Model	9	12,869.1	1429.89	44.30	0.000				
Linear	3	09238.3	3079.42	95.40	0.000				
Square	3	02258.7	0752.89	23.32	0.001				
Interaction	3	01372.1	0457.37	14.17	0.002				
Error	7	00225.9	0032.28						
Lack of fit	5	00216.6	0043.32	09.29	0.100				
Pure error	2	00009.3	0004.66						
Total	16	13,095.0							

Table 5 Summary of significance test results for sand mould properties

Output	Coefficient of correlations		Parameters	
	All terms	Exclude insignificant terms	Significant terms	Insignificant terms
GE	0.9393	0.8613	A, C, AA, BB, CC, AB, AC, BC	B
CS	0.9952	0.9821	A, B, C, AA, BB, CC, AB, AC	BC
CP	0.9877	0.9720	A, B, C, AA, AB, AC	BB, CC, BC
MH	0.9744	0.9414	A, B, C, AA, AB, AC	BB, CC, BC
P	0.9827	0.9606	A, B, AA, AB, AC	C, BB, CC, BC

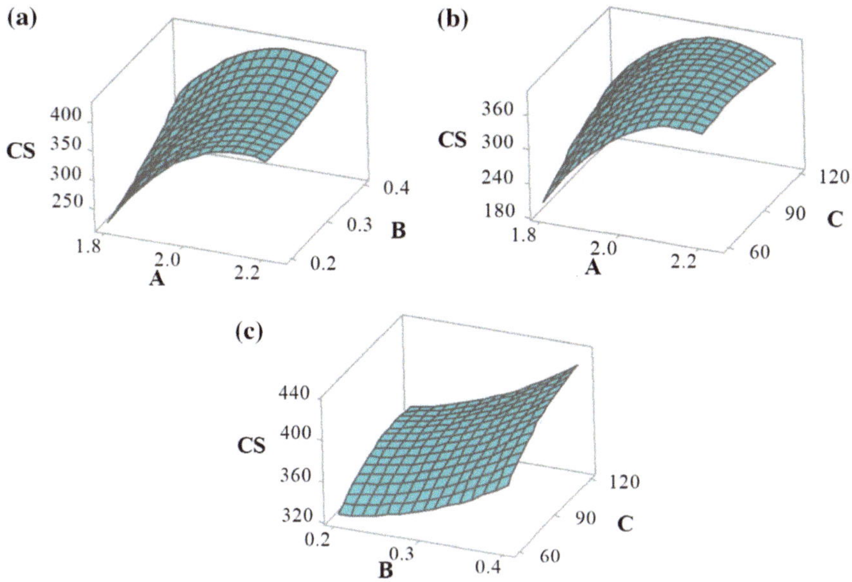

Fig. 2 3D surface graphs of compression strength with: **a** percent of resin and percent of hardener, **b** percent of hardener and curing time and **c** percent of hardener and curing time

for the compression strength is presented in Table 4. All linear, corresponding square and combined interaction terms (excluding the interaction among percent of hardener and curing time) are found to be significant for the response, CS (refer Table 5). Insignificant term depict there is no significant change in the output value when the independent variables are varied simultaneously within their operating levels. The *P*-values of all square terms are found to be significant (as their corresponding *P*-value is found to be less than 0.05), indicating all sand mould variables (that is, percent of resin, percent of hardener and curing time) are found to have non-linear relation with the response, CS. The results of statistical tests are found to be in line with the 3D surface plots (refer Fig. 2).

Figure 2 shows the 3-dimensional response surface plots drawn to know the impact of experimental (input) factors on the compression strength, when two variables are varied simultaneously within their operating range and keeping the remaining parameters at fixed center level. The observations made from the surface plots are as follows:

1. Figure 2a shows the interaction factors effect between the percent of resin and percent of hardener on the response, CS. It is observed that, the compression strength tends to increase with percentage of resin and hardener. The results showed that the resin tends to contribute more in comparison with hardener for the CS. Lower resin quantity might not be sufficient enough to coat the sand grains, will not develop strong bonding action between the sand grains. Further,

low quantity of hardener might not be sufficient enough to stimulate the available resin. The results are in line with the experiments conducted earlier by Bargaoui et al. [21].
2. High values of compression strength are seen with the increased values of percent of resin (refer Fig. 2b). Further, CS is found to have a negligible impact when the curing time is varied from low to high values. This implies high values of hardener content support polymerization that develops strong bonding action between the molecules of resin to coat on sand grains.
3. CS is found to increase linearly with an increase in hardener and curing time simultaneously. This indicates high quantity of hardener might not provide strong bonding strength with low curing time. Increase in curing time will provide sufficient time for the resin to undergo polymerization and develop strong bonding action between the sand grains which improves the mould compression strength.

The multiple correlation coefficient established for the compression strength is found to be close to 1, which indicates the model fits to the assumed regression equation with good precision. The ANOVA result shows that, the model developed for the response, CS is statistically adequate. The combined effect of linear, square and 2-term interactions and lack-of-fit was found significant. Further, the model needs to be tested for its accuracy in prediction by utilizing test cases with randomized combination of variable parameter values. However, it is to be noted that, the variable parameters should be within their operating range.

4.1.2 Permeability

The relationship of the response permeability with the sand mould variables (percent of resin, curing time and percent of hardener) is represented mathematically as follows,

$$P = 3178 - 2653A - 342B - 4.320C + 566.0A^2 - 696B^2 + 0.00732C^2 + 367AB + 1.658AC - 1.433BC \quad (2)$$

The significance of moulding sand variables, their curvature, and two-term interactions are tested at 95% confidence interval. The obtained significance test results are discussed below (refer Table 5).

1. The variable, curing time (that is, C) is not having significant contribution towards this response.
2. The quadratic terms of variables, namely, percent of hardener and curing time are not significant towards the response, permeability. This indicates that, the existence of strong dependent linear relationship of these parameters with permeability.
3. Although percent of hardener alone has showed a significant impact, their interaction with curing time is found insignificant. This indicates the permeability

does not depend much on the interaction of curing time and percent of hardener (i.e. BC).

The combined effect of all linear (A, B, C), corresponding square (A^2, B^2 and C^2), and combined two-term interaction (AB, AC, and BC) term effects is found to be significant at 95% confidence level (refer Table 4). Excluding insignificant terms will result in imprecise input-output relationship and might reduce the prediction accuracy. The multiple correlation coefficient obtained for the response permeability is found equal to 0.9827 (refer Table 5). Hence, the models are statistically adequate to make use for prediction of permeability for known set of sand mould variables.

4.1.3 Mould Hardness

Mould hardness, expressed as a mathematical non-linear function of percent of resin, percent of hardener and curing time is shown below-

$$\begin{aligned} \mathrm{MH} = & -334.0 + 322.0\mathrm{A} + 213.0\mathrm{B} + 0.751\mathrm{C} - 65.4\mathrm{A}^2 - 91.8\mathrm{B}^2 - 0.000964\mathrm{C}^2 \\ & - 71.2\mathrm{AB} - 0.2542\mathrm{AC} + 0.017\mathrm{BC} \end{aligned} \quad (3)$$

The result of significance test for the response-mould hardness shows that, the combined effect of all linear, quadratic, two-factor terms and lack-of-fit are statistically significant at 95% confidence level (refer Table 4). The terms (i.e. B^2, and C^2) are found insignificant, indicating percentage of hardener and curing time have strong linear relationship with the response, mould hardness. Although hardener and curing time are found to have significant contribution, its interaction (i.e. BC) is insignificant towards this response. Percent of resin has maximum contribution, followed by hardener and curing time towards the response, mould hardness. The model found to be statistically adequate with good fit of response surface and resulted in a better correlation coefficient value equal to 0.9744 (refer Table 5). Therefore, the model can be used to make mould hardness prediction for known combination of sand mould variables.

4.1.4 Gas Evolution

The mathematical model established between gas evolution and sand mould variables (i.e. percent of resin, percent of hardener and curing time) using experimental data is shown below,

$$\begin{aligned} \mathrm{GE} = & \ 59.1 - 49.7\mathrm{A} - 81.6\mathrm{B} + 0.2026\mathrm{C} + 12.45\mathrm{A}^2 + 44.8\mathrm{B}^2 - 0.000669\mathrm{C}^2 \\ & + 21.81\mathrm{AB} - 0.0544\mathrm{AC} + 0.1112\mathrm{BC} \end{aligned} \quad (4)$$

Table 4 shows the significance test results obtained for GE. The percent of hardener is not found to be significant, because their *P*-value is more than 0.05. The analysis of variance values of full quadratic (linear, square, and two term interaction) terms for the gas evolution is shown in Table 5. The *P*-values of full quadratic terms are found to be lower than 0.05, indicating good fit of response surface for GE. Further, the coefficient of determination value obtained for gas evolution is found to be equal to 0.9393, indicating the mathematical models are capable to predict accurately the response, GE. Excluding non-significant terms from the derived response Eq. (4), will make the lack-of-fit significant.

4.1.5 Collapsibility

The second order response surface model between the moulding sand variables and collapsibility is expressed as follows:

$$\begin{aligned} CP = & -6615 + 5993A + 1630B + 7.48C - 1290A^2 + 934B^2 - 0.01034C^2 \\ & - 1037AB - 2.880AC + 1.85BC \end{aligned} \quad (5)$$

The significance of linear factors (i.e. sand mould variables), their nonlinearity (i.e. curvature), and two term interactions are evaluated for the confidence level of 95%. It is important to note that the collapsibility and mould hardness identified similar significant and insignificant terms (refer Table 5). This might have happened due to the existence of strong dependency between the mould hardness and collapsibility. The coefficient of determination for the response, collapsibility was found equal to 0.9877 (refer Table 5). Since, the coefficient of correlation is close to 1, prediction ability by the response equation will be close the actual values of collapsibility.

4.1.6 Prediction Test for the Developed Non-linear Models

The discussion from previous section indicates, all non-linear regression models developed for the responses, namely CS, P, MH, GE and CP are statistically adequate at 95% confidence level. These models are tested towards their prediction capability by conducting 14 experiments (test cases), which will be treated as target values. It is to be noted that these experiments are conducted with different set of variable combination generated at random. The values of variables are generated at random within their operating range (Appendix 1). The percent deviation in predicting the sand mould properties (i.e. CS, P, MH, GE and CP) for 14 test cases are presented in Appendices 2 and 3. The Percentage deviation in predicting the response value is found to vary in the ranges between −6.58 and +6.26% for CS, −9.53 and +5.34% for CP, −4.05 and +4.29% for GE, −4.86 and +3.29% for MH, and −8.38 and + 6.18% for P (Appendices 2 and 3). It is important to mention that, the % deviation is found to vary on both positive and negative sides and vary within the acceptable

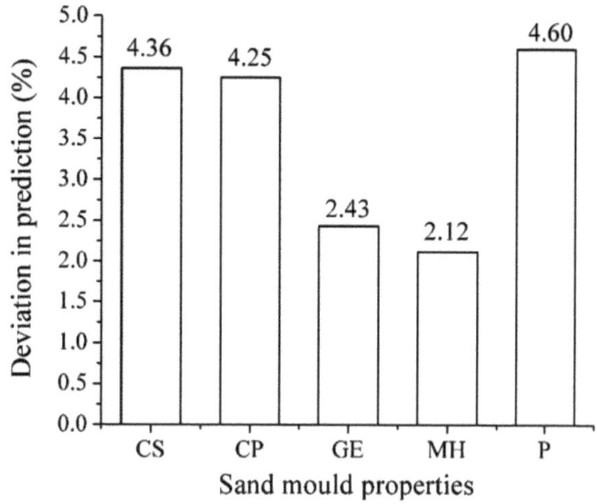

Fig. 3 Mean absolute percent error in prediction of sand mould properties

range for all the responses (refer Appendices 2 and 3). This shows that, the model has accurately captured the process physics, mechanics and dynamics. Further, the mean absolute percent deviation in prediction for CS, CP, GE, MH and P is found equal to 4.36, 4.25, 2.43, 2.12 and 4.6%, respectively (refer Fig. 3). The mean absolute percent deviation in prediction of all sand mould properties (i.e. CS, CP, GE, MH and P) is found equal to 3.55%. This depicts the developed non-linear models can be used by any novice user to predict the sand mould properties without the requirement of prior knowledge and conducting trial experiments.

4.2 Multi-response Optimization

Accurate control of moulding sand properties with conflicting requirements is a tedious task in foundry industries. A situation might arise in shop floor such that a set of sand mould variables might results in better permeability, but not offer the desired strength and hardness of moulds. This is due to the fact that, the response permeability has inverse relation with mould hardness and compression strength. Multi-objective optimization would solve this complex situation by determining an optimal set of sand mould variables (that is, percent of resin, percent of hardener and curing time) for the conflicting requirements in sand mould properties (that is, minimize: GE and CP, and maximize: CS, MH and P). The upper and lower constrained values of sand mould variables could define the three-dimensional solution spaces, which will help optimization tools (DFA, GA, PSO, and TLBO) to conduct search for the best sand mould properties. Optimal sand moulding properties for the conflicting objective functions (simultaneous, maximization and minimization) in alphaset sand moulding system will need a suitable mathematical formulation. Weight average method is employed to convert multiple conflicting objective functions to a single objective function either for maximization or minimization [14, 34]. The present work require optimization of five conflicting objective functions. Six different case studies are considered by assigning equal importance (i.e. 20 wt%) to all outputs and maximum importance to one output at a time (that is, 60%), with the rest at low and equal weights (that is, 10%). The optimal search is conducted. Accurate control of moulding sand properties with conflicting requirements is treated as a tedious task in shop floor foundry. A situation might arise in shop floor such that a set of sand mould variables might results in better permeability, but not offer the desired strength and hardness of moulds due to permeability pose inverse relation with mould hardness and compression strength. Multi-objective optimization would solve this complex situation by determining optimal set of sand mould variables (that is, percent of resin, percent of hardener and curing time) for the conflicting requirements in sand mould properties (that is, minimize: GE and CP, and maximize: CS, MH and P). The upper and lower constrained values of sand mould variables could define the three-dimensional solution spaces, which help optimization tools (DFA, GA, PSO, and TLBO) to conduct optimal search for best sand mould properties. Optimal sand moulding properties for the conflicting objective functions (simultaneous, maximization and minimization) in alphaset sand moulding system require suitable mathematical formulation. Weight average method is employed to convert multiple conflicting objective functions to single function either for maximization or minimization [14, 34]. The present work require optimization of five conflicting objective functions, six different case studies are considered after assigning equal importance (i.e. 20 wt%) to all outputs and maximum importance to single output (that is, 60%), with the rest at low and equal weights (that is, 10%). Optimization tools are used to obtain the best set of mould properties. Optimization of mathematically formulated weighted objective (output) function for maximization is discussed as follows,

Output function (Y_1) = CS
Output function (Y_2) = P
Output function (Y_3) = MH
Output function (Y_4) = 1/GE
Output function (Y_5) = 1/CP

$$Maximize\ (Y) = W_1Y_1 + W_2Y_2 + W_3Y_3 + W_4Y_4 + W_5Y_5$$

Terms W_1Y_1, W_2Y_2, W_3Y_3, W_4Y_4, and W_5Y_5 are the weight fraction combination for the objective function CS, P, MH, GE and CP, respectively. Weight factors (W_1–W_5) combination are selected such that their cumulative value must be equal to one. In Alphaset sand moulding process, the sand mould properties are imposed by parameter upper and lower bound constraints which cover percent of resin, percent of hardener and curing time. These variable constraints are listed in Table 6.

Table 6 Upper and lower bound of constrained variables

Parameters	Lower bound	Upper bound
Percent of resin, %	1.8	2.2
Percent of hardener, %	0.2	0.4
Curing time, s	60	120

4.2.1 Desirability Function Approach (DFA)

In 1980, Derringer and Suich had proposed the desirability function approach for multi-response optimization [37]. Reduced gradient approach was employed to locate the optimal solutions, which initiate with multiple solution and end with global solution (i.e. highest desirability) [38]. The desirability (D) value could vary between the ranges of 0 and 1. The solutions are completely acceptable (i.e. output function value is perfectly the target or global value) when the D = 1 or close to 1. The present work is focussed to optimize the conflicting objective functions which have both maximizing and minimizing the individual desirability functions.

The responses (CS, MH, and P) are of maximizing type and the individual desirability function is presented by Y_{CS}, Y_P, and Y_{MH}.

$$y_{CS} = \frac{CS - CS_{\min}}{CS_{\max} - CS_{\min}}, \quad y_P = \frac{P - P_{\min}}{P_{\max} - P_{\min}}, \quad \text{and} \quad y_{MH} = \frac{MH - MH_{\min}}{MH_{\max} - MH_{\min}}$$

where,

P_{max} and P_{min} is the maximum and minimum value of P
CS and CS_{min} is the maximum and minimum value of CS
MH_{max} MH_{min} is the maximum and minimum value of MH.

The responses (GE, and CP) are of minimizing type and the individual desirability function is presented by Y_{GE}, and Y_{CP}.

$$y_{GE} = \frac{GE\text{max} - GE}{GE\text{max} - GE_{\text{min}}} \quad \text{and} \quad y_{CP} = \frac{CP\text{max} - CP}{CP\text{max} - CP_{\text{min}}}$$

where,

GE_{max} and GE_{min} is the maximum and minimum value of GE
CP_{max} and CP_{min} is the maximum and minimum value of CP.

For multi-objective functions the highest composite desirability value obtained from six different case studies is treated as an optimal choice for Alphaset sand moulding process. The single composite desirability value, satisfying all conflicting requirements in sand mould properties is computed as shown below

$$D_0 = \sqrt[5]{y_{CS}^{w_1} \times y_P^{w_2} \times y_{MH}^{w_3} \times y_{GE}^{w_4} \times y_{CP}^{w_5}} \tag{6}$$

4.2.2 Genetic Algorithm (GA)

In 1975, Holland introduced the concept of genetic algorithm which mimic the natural selection of living organisms based on Charles Darwin theory of survival of fittest [14, 34]. The process starts with initialization of genetic operators (mutation, crossover and selection) and generation of population to produce local solutions. To obtain global solutions the decision on optimum selection of genetic operators are of primary importance. There are no universal standards or methods established to obtain the global solutions for optimum choice of GA parameters (i.e. genetic operators, population size and generation or iteration number). Thereby, parameter study is conducted to ensure the highest desirability (D_0) values correspond to GA parameters. Tournament selection method is employed to rank the obtained solutions by balancing the diversity of new population and enhance the current solution. The optimum GA parameters correspond to highest desirability (D_0) value is decided by conducting parameter study is presented in Table 7.

4.2.3 Particle Swarm Optimization: PSO

In 1995, Dr. Eberhart and Dr. Kennedy presented the concept of particle swarm optimization at the Congress on Evolutionary Computation (Kennedy and Eberhart 1995). PSO imitate the cooperation among individuals and in the team utilizing swarm intelligence and share experiences from one generation to other. PSO pose technical advantage over GA, as PSO requires few tuning (inertia weight, swarm size, mutation rate and generation) parameters [14, 34], and do not require sorting of fitness values and the solution lead to fast convergence. Mutation operator is introduced to simple PSO to enlarge the search space to avoid local minima, if any [14]. Systematic study results of optimization of particle swarm optimization parameters for highest composite desirability (D_0) value correspond to the sand moulding properties and sand mould variables are presented in Table 7.

Table 7 Parameter study results of GA and PSO

GA parameter study			PSO parameter study		
Parameters	Levels	Optimum value	Parameters	Levels	Optimum value
Cross over rate	0.3–1.0	0.5	Inertia weight	0–1	0.55
Mutation rate	0.03–0.3	0.15	Mutation rate	0.03–0.3	0.09
Population size	20–180	150	Swarm size	20–180	100
Generations	20–1000	500	Generations	20–1000	200

4.2.4 Teacher Learning Based Optimization (TLBO)

Rao introduced the concept of teaching learning-based optimization algorithm. TLBO is also a population-based algorithm [30], wherein the group of learners or class of learners is treated as populations. TLBO algorithm is considered to be more efficient for solving the non-linear optimization problem, and do not require algorithm specific tuning parameters. Thus, TLBO algorithm converge solutions at faster rate and avoids local minima solution due to improper choice of tuning parameters. The algorithm works in two phases (i.e. teacher and learner phase) to locate optimal solutions. In teacher phase, teacher is the highly trained professional who motivate the students to acquire greater knowledge and always focussed to improve the mean result of a class. In learner phase, apart from acquire knowledge in teacher phase, the learner's mutual interaction could also help to improve the mean result of a class. Although there are no algorithm specific tuning parameters, the size of population and generations are required to be optimized in TLBO. TLBO parameters are optimized for highest desirability (D_O) value as below:

Number of population = 60
Number of generation = 100.

4.2.5 Comparison of Performance of Optimization Models: DFA, GA, PSO, and TLBO

Table 8 shows the different optimization tools (DFA, GA, PSO, and TLBO) used to locate extreme values of sand mould properties and corresponding sand mould variables for six case studies. The values with highest global desirability will define the best sand mould conditions (percent of resin, percent of hardener, and curing time) and properties. The optimum set of sand mould properties, located with the help of fine-tuned GA, PSO and TLBO parameters is used to determine the global desirability value for six different case studies. The choice of best sand moulding conditions from six different case studies is obtained with the help of highest desirability (D_O) value. The composite desirability value obtained for case 1–6 are found to be {0.8826, 0.8873, 0.8881, 0.8947}, {0.9073, 0.9076, 0.9109, 0.9122}, {0.9033, 0.9098, 0.9185, 0.9152}, {0.9199, 0.9138, 0.9275, 0.9297}, {0.9099, 0.9146, 0.9351, 0.9382}, and {0.9031, 0.9059, 0.9117, 0.9096} for DFA, GA, PSO, and TLBO, respectively (refer Table 8). Important to note that, TLBO outperformed PSO, DFA, and GA to locate the highest composite desirability (D_o) value. TLBO determined case 5 (that is, highest importance assigned to GE) is recommended as an optimum choice for Alphaset-sand moulding system, as their corresponding desirability function value found to be maximum compared to other cases studied. Further, TLBO optimizes the sand mould properties with less population size and generation {60 and 100} number as compared to GA {150, 500} and PSO {100, 200} (refer Table 7). Further, TLBO does not require algorithm specific tuning parameters unlike in GA (crossover, and mutation) and PSO (inertia weight, social and cognitive leader) and lead to faster

Table 8 Optimum sand moulding conditions for multiple outputs with different combination of weight factors via DFA, GA, PSO, and TLBO

Case studies	Models	Desirability (D_0)	Responses							
			A	B	C	CS, KPa	P	MH	GE, ml/gm	CP, KPa
Case 1: equal importance to all outputs (W_1, W_2, W_3, W_4 and $W_5 = 0.2$)	DFA	0.8826	1.87	0.27	79.3	283.6	157.7	70.44	8.09	204.7
	GA	0.8873	1.88	0.24	114.4	312.3	154.3	72.85	8.30	218.5
	PSO	0.8881	1.83	0.32	066.6	252.6	169.0	68.42	7.35	182.7
	TLBO	0.8947	1.86	0.35	60.42	277.7	157.8	69.22	6.90	207.5
Case 2: highest importance to CS ($W_1 = 0.6$, W_2, W_3, W_4 and $W_5 = 0.1$)	DFA	0.9073	2.17	0.23	118.7	363.5	131.8	75.99	8.00	273.7
	GA	0.9076	1.96	0.27	112.5	360.4	132.4	75.76	8.00	266.7
	PSO	0.9109	2.04	0.22	119.3	368.7	131.5	75.73	7.85	274.2
	TLBO	0.9122	2.04	0.27	119.4	375.9	127.5	76.75	7.75	281.5
Case 3: highest importance to P ($W_2 = 0.6$, W_1, W_3, W_4 and $W_5 = 0.1$)	DFA	0.9033	1.8	0.23	83.6	222.42	184.0	66.59	8.82	144.9
	GA	0.9098	1.83	0.32	66.6	252.6	169.0	68.42	7.35	182.7
	PSO	0.9185	1.81	0.32	60.8	219.9	182.4	66.39	7.06	155.7
	TLBO	0.9152	1.82	0.29	63.9	230.8	178.2	67.01	7.4	163.4
Case 4: highest importance to MH ($W_3 = 0.6$, W_1, W_2, W_4 and $W_5 = 0.1$)	DFA	0.9199	2.14	0.31	118.7	382.6	120.9	77.23	8.14	289.9
	GA	0.9138	1.91	0.33	91.3	348.93	130.5	74.57	8.07	262.3
	PSO	0.9275	1.91	0.32	118.6	363.02	130.4	76.29	7.92	269.5
	TLBO	0.9297	1.94	0.29	119.4	359.3	134.1	76.06	7.82	264.8
Case 5: highest importance to GE ($W_4 = 0.6$, W_1, W_2, W_3 and $W_5 = 0.1$)	DFA	0.9099	1.87	0.36	72.4	315.1	142.1	71.74	7.51	237.5
	GA	0.9146	1.98	0.30	61.3	332.1	135.2	72.28	7.34	253.4
	PSO	0.9351	1.86	0.37	61.7	296.2	149.2	69.95	6.93	224.1
	TLBO	0.9382	1.84	0.38	60.8	285.7	153.2	69.21	6.84	215.8
Case 6: highest importance to CP ($W_5 = 0.6$, W_1, W_2, W_3 and $W_4 = 0.1$)	DFA	0.9031	1.84	0.24	62.3	226.95	178.8	65.88	7.85	159
	GA	0.9059	1.8	0.28	81.2	239.57	176.4	68.4	8.21	164.3
	PSO	0.9117	1.81	0.32	60.8	219.92	182.4	66.39	7.06	155.7
	TLBO	0.9096	1.81	0.29	69.9	226.69	180.4	67.17	7.72	157.4

Table 9 Validate the optimization models with experimental sand mould conditions

D_0	Condition	Sand mould variables (inputs)			Sand mould properties (outputs)				
		A	B	C	CS, KPa	P	MH	GE, ml/gm	CP, KPa
0.9382	Optimized	1.84	0.38	60.8	285.7	153.2	69.21	6.84	215.8
0.7861	Normal	2.20	0.40	120	388.9	106.6	75.60	9.28	285.2

convergence. Confirmation experiments are conducted for the determined optimal sand mould conditions with highest composite desirability value of Case 5 by TLBO. TLBO results are compared with experimental values to confirm their practical utility. The results are found to be more useful for a foundry personal as their deviation is found to be less than 8% with enhanced sand mould properties. The experiments are also conducted for the lowest desirability value obtained from Table 9 (that is, Experiment No. 2) for the normal sand moulding condition. This experiment is done to know the importance of optimization methods in achieving the casting quality. Important to note that, although normal sand moulding condition (refer Table 9) provide better strength, but affects the mould permeability, gas evolution and collapsibility. Low permeable moulds and high gas evolution increase the mould pressure which create inadequate space for the generated gases during metal pouring and will result in gas porosity in the castings [13]. Further, poor collapsibility requires additional equipment to break the moulds, which might induce the residual stresses and hot tear defects in castings [11].

5 Casting Quality Assessments for Different Sand Mould (Optimized and Normal) Conditions

The Alphaset resin bonded sand mould prepared with optimized and normal sand mould condition is used to cast the automotive bushing (refer Table 10). The castings are subjected to different tests according to reference standards to evaluate the quality characteristics. LM20 molten metal is poured to the prepared mould cavity. The mould cavity is prepared with the help of aluminium pattern.

The casting, automotive bushing part is tested for different quality (density, SR, YS, UTS, SDAS, and BHN) characteristics (refer Fig. 4a) and evaluated the difference in quality for the castings obtained from optimized and normal moulding sand conditions. The outer casting surface structure is evaluated with the help of Mitutoyo Surftest SJ301. Gas porosity was observed on the tensile fracture surface for the castings prepared by normal sand mould condition (refer Fig. 4d).

Homogeneous texture with increased amount of fine dimples on the fractured intergranular surface is observed on tensile specimens (refer Fig. 4e). This indicates relatively large amount of plastic deformation occurs before fracture (refer

Table 10 Casting quality characteristics under different moulding conditions

Casting quality characteristics	Notation	Testing Standards	Optimized condition	Normal condition
Surface roughness, μm	SR	JIS 2001	4.88	5.63
Yield strength, MPa	YS	ASTM E8	112	104
Ultimate tensile strength, MPa	UTS	ASTM E8	183	158
Hardness	BHN	ASTM E10	61	58
Density, g/cm3	ρ	Archimedes	2.62	2.59
Secondary dendrite arm spacing, μm	SDAS	ASTM E112	49.0	55.0

Fig. 4e). The resulted microstructure is dendritic structure with fine α-aluminium grains. The refined silicon particles have appeared in the matrix between the dendrites and resulted in a low value of secondary dendrite arm spacing (refer Fig. 4c).

Segregations are formed between the dendrites for the castings obtained by normal sand mould condition (refer Fig. 4b). The surface finish, dendritic structure, and fracture mechanism obtained for highest desirability value (i.e. 0.9382) is found to be better as compared to low values of global desirability (i.e. 0.7861). Three replicates are considered for each casting condition and the average value of different quality characteristics (i.e. 9 SR, 3 YS and UTS, 9 BHN, and 3 SDAS) at different section on the casting samples are presented in Table 10. The casting quality obtained for the optimized sand mould condition experiment is found to be better compared to that obtained with the normal sand mould condition (refer Table 10). That is, the casting quality made in mould with highest desirability value is found to be better than the casting made in mould with lower desirability value.

6 Conclusion

The main objective of the present research work is to apply statistical modelling and optimization tools in Alpha-set moulding sand system to improve the quality of mould and thereby the quality of casting. The research work will help the foundry personnel to produce good quality castings with lower cost. Modelling has been carried out by conducting experiments, as per CCD matrix with different combination of sand mould variables (i.e. percent of resin, percent of hardener and curing time). The responses considered in the study include sand mould properties, namely CS, P, MH, GE, and CP. Statistical (significance, analysis of variance, and prediction) tests, and 3-dimensional response surface analysis are carried out to know the behaviour of

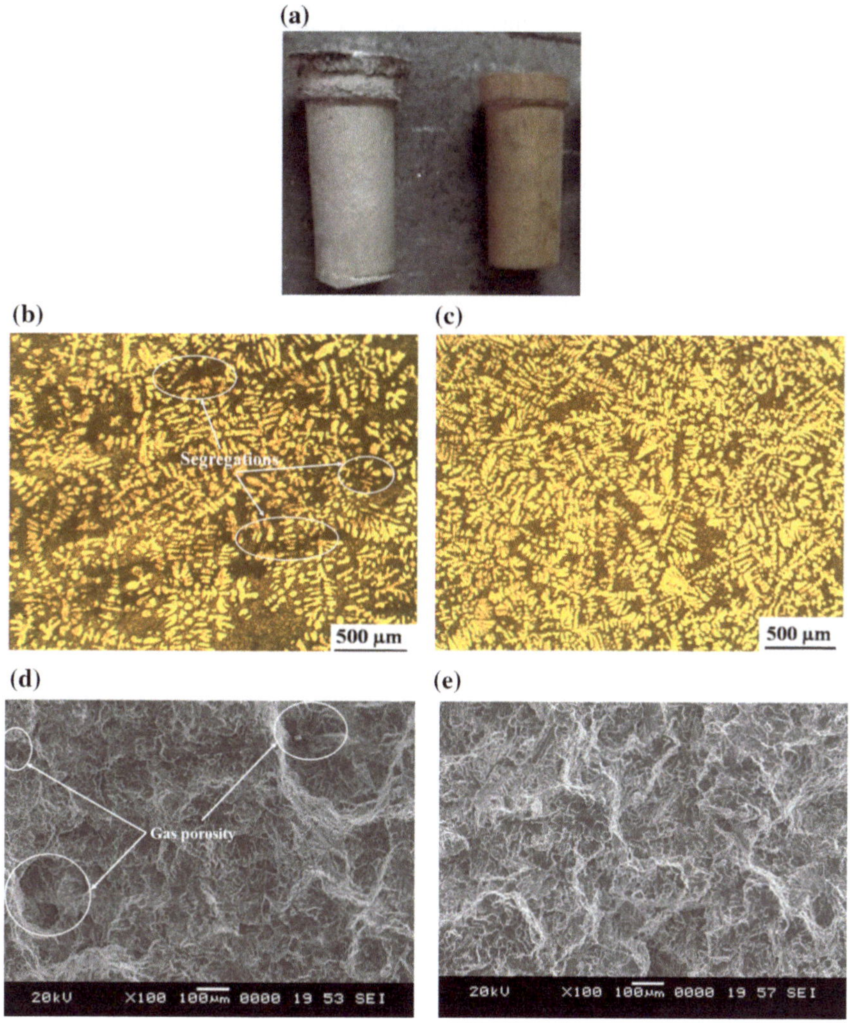

Fig. 4 **a** Photograph of cast samples, dendrite structure of **b** normal sand mould and **c** optimized sand mould, fracture surface of **d** normal sand mould and **e** optimized sand mould

variables, physics and mechanics of a process accurately. Optimization tools, namely DFA, GA, PSO and TLBO are applied to optimize the multiple outputs (maximize: CS, P, and MH, and minimize: GE, and CP). The following conclusions are drawn from the present work:

1. All linear factors (i.e. percent of resin, percent of hardener and curing time) are found to have significant contribution towards all sand mould properties (excluding, curing time for permeability).

2. The quadratic effects of percent of hardener and curing time is found insignificant (as P-values > 0.05) for the responses, namely collapsibility, mould hardness and permeability. This indicates CS, MH and P relationship with percent of hardener and curing time is linear in nature. Important to note that, the interaction among percent of hardener and curing time is insignificant for all the sand mould properties (excluding, GE).
3. Non-linear models developed for all sand mould properties are found to be statistically adequate with good fit of response surfaces. The prediction accuracy of all non-linear models is found to be less than 5%. The better prediction accuracy might be due to the fact that, model captured the physics, mechanics and dynamics of a process accurately. Further, the developed models help any novice user to predict the sand mould properties without conducting the practical experiments.
4. The sand mould properties are complex, non-linear and conflicting (maximize: CS, P, and MH, and minimize: GE and CP) in nature. The optimum solutions are many due to multiple objective functions with conflicting requirements. Weight based method is used to determine the optimal solutions after assigning different combination of weight fractions (importance) to each individual output. The highest desirability (D_O) value corresponds to the different cases is studied and treated as an optimal condition for sand mould properties. Confirmation experiments revealed that, the optimized sand mould properties will produce the casting with better quality characteristics compared to normal sand moulding conditions.

Appendix 1: Test Case Data for Sand Moulding Variables and Moulding Sand Properties

Test no.	Moulding sand variables			Moulding sand properties				
	% of resin	% of hardener	Curing time	CS, KPa	P	MH	CP, KPa	GE, ml/gm
1	1.85	0.20	064	212.4	187.3	62.04	142.2	8.40
2	2.20	0.25	118	376.8	138.6	76.74	256.8	8.34
3	1.90	0.30	106	360.7	134.3	73.28	258.9	7.82
4	1.95	0.35	074	367.3	120.2	71.20	256.3	7.73
5	1.80	0.35	089	271.8	159.9	72.48	216.0	8.02
6	2.10	0.30	102	414.0	108.4	79.84	290.4	8.72
7	1.95	0.25	096	332.2	142.8	76.23	260.6	8.21
8	2.20	0.40	063	381.6	100.7	74.03	286.7	9.45
9	2.15	0.20	074	361.1	121.3	75.32	292.4	9.20
10	2.05	0.35	096	428.8	99.8	78.13	289.7	8.63
11	1.85	0.30	088	276.3	164.3	72.38	218.9	8.04
12	2.20	0.25	112	350.8	132.4	75.30	280.1	8.80
13	1.80	0.25	092	247.8	182.7	65.23	151.5	8.38
14	2.05	0.35	078	402.5	101.4	76.40	278.7	8.08

Appendix 2: Summary Results of Model Predicted Test Cases of Sand Mould Properties (CS, P, MH, GE and CP)

Test no.	Compression strength, KPa				Permeability			
	Exp. value	Model prediction	Deviation (%)	Absolute deviation (%)	Exp. value	Model prediction	Deviation (%)	Absolute deviation (%)
1	212.4	225.17	−6.01	−6.01	187.3	178.10	4.91	4.91
2	376.8	355.63	5.62	5.62	138.6	134.00	3.32	3.32
3	360.7	342.85	4.95	4.95	134.3	137.19	−2.15	2.15
4	367.3	356.35	2.98	2.98	120.2	124.92	−3.93	3.93
5	271.8	282.86	−4.07	−4.07	159.9	157.17	1.71	1.71
6	414.0	389.05	6.03	6.03	108.4	115.54	−6.59	6.59
7	332.2	343.86	−3.51	−3.51	142.8	135.50	5.11	5.11
8	381.6	390.71	−2.39	−2.39	100.7	106.22	−5.48	5.48
9	361.1	374.83	−3.80	−3.80	121.3	114.94	5.24	5.24
10	428.8	401.94	6.26	6.26	99.8	107.21	−7.43	7.43
11	276.3	294.49	−6.58	−6.58	164.3	154.15	6.18	6.18
12	350.8	360.63	−2.80	−2.80	132.4	130.08	1.75	1.75
13	247.8	239.59	3.31	3.31	182.7	178.71	2.18	2.18
14	402.5	391.30	2.78	2.78	101.4	109.90	−8.38	8.38
	Mould hardness				Collapsibility			
1	62.04	64.69	−4.27	4.27	142.2	155.8	−9.53	9.53
2	76.74	75.92	1.06	1.06	256.8	267.1	−4.02	4.02
3	73.28	74.88	−2.18	2.18	258.9	252.3	2.56	2.56
4	71.20	73.98	−3.91	3.91	256.3	272.5	−6.33	6.33
5	72.48	71.16	1.82	1.82	216.0	204.5	5.34	5.34
6	79.84	77.21	3.29	3.29	290.4	296.2	−2.01	2.01
7	76.23	74.05	2.86	2.86	260.6	254.5	2.34	2.34
8	74.03	74.40	−0.50	0.50	286.7	292.5	−2.03	2.03
9	75.32	74.40	1.22	1.22	292.4	290.4	0.67	0.67
10	78.13	77.23	1.15	1.15	289.7	308.5	−6.47	6.47
11	72.38	71.68	0.97	0.97	218.9	212.7	2.84	2.84
12	75.30	76.08	−1.03	1.03	280.1	271.7	2.98	2.98
13	65.23	68.40	−4.87	4.87	151.5	158.3	−4.48	4.48
14	76.40	76.01	0.52	0.52	278.7	300.8	−7.94	7.94

Appendix 3: Summary Results of the Test Cases for the Responses—GE

Test no.	Exp. value	Model prediction	Deviation (%)	Absolute deviation (%)
1	8.40	8.52	−1.37	1.37
2	8.34	8.16	2.12	2.12
3	7.82	8.14	−4.05	4.05
4	7.73	7.70	0.41	0.41
5	8.02	8.13	−1.34	1.34
6	8.72	8.38	3.88	3.88
7	8.21	8.33	−1.43	1.43
8	9.45	9.11	3.60	3.60
9	9.20	8.97	2.55	2.55
10	8.63	8.43	2.35	2.35
11	8.04	8.15	−1.36	1.36
12	8.80	8.42	4.29	4.29
13	8.38	8.72	−4.04	4.04
14	8.08	8.18	−1.27	1.27

References

1. Holtzer M, Dańko R, Żymankowska-Kumon S (2014) The state of art and foresight of world's casting production. Metalurgija Metall 53:697–700
2. Chojecki A, Mocek J (2011) Gas pressure in sand mould poured with cast iron. Arch Foundry Eng 11:9–14
3. Alonso-Santurde R, Coz A, Quijorna N, Viguri JR, Andres A (2010) Valorization of foundry sand in clay bricks at industrial scale. J Ind Ecol 14(2):217–230
4. Bobrowski A, Grabowska B (2012) The impact of temperature on furan resin and binder structure. Metall Foundry Eng 38(1):73–80
5. Holtzer M, Zymankowska-Kumon S, Bobrowski A, Kmita A, Dańko R (2015) Influence of the reclaim addition to the moulding sand matrix obtained in the ALPHASET technology on the emission of gases—comparison with moulding sand with furfuryl resin. Arch Foundry Eng 15(1):121–125
6. Vasková I, Smolková M, Malik J, Eperješi Š (2008) Experience in forming and core mixtures by Alphaset technology. Arch Foundry Eng 8(2):141–144
7. Mocek J, Samsonowicz J (2011) Changes of gas pressure in sand mould during cast iron pouring. Arch Foundry Eng 11(4):87–92
8. Holtzer M, Dańko R, Górny M (2016) Influence of furfuryl moulding sand on flake graphite formation in surface layer of ductile iron castings. Int J Cast Met Res 29(1–2):17–25
9. Major-Gabrys K, StM Dobosz, Jakubski J (2011) The estimation of harmfulness for environment of moulding sand with biopolymer binder based on polylactide. Arch Foundry Eng 11:69–72

10. Izdebska-Szanda I, Szanda M, Matuszewski S (2011) Technological and ecological studies of moulding sands with new inorganic binders for casting of non-ferrous metal alloys. Arch Foundry Eng 11(1):43–48
11. Roy T (2013) Analysis of casting defects in foundry by computerised simulations (CAE)—a new approach along with some industrial case studies. In: Transactions of 61st Indian Foundry Congress 2013, pp 1–9
12. Parappagoudar MB, Pratihar DK, Datta GL (2006) Non-linear modelling using central composite design to predict green sand mould properties. Proc IMechE Part-B J. Eng Manuf 221, 881–895
13. Reddy NS, Yong-Hyun B, Seong-Gyeong K, Young HB (2014) Estimation of permeability of green sand mould by performing sensitivity analysis on neural networks model. J Korea Foundry Soc 34(3):107–111
14. Surekha B, Kaushik LK, Panduy AK, Vundavilli PR, Parappagoudar MB (2012) Multi-objective optimization of green sand mould system using evolutionary algorithms. Int J Adv Manuf Technol 58(1–4):9–17
15. Nastac Laurentiu, Jia Shian, Nastac Mihaela N, Wood Robert (2016) Numerical modelling of the gas evolution in furan binder-silica sand mold castings. Int J Cast Met Res 29(4):194–201
16. Holtzer M, Bobrowski A, Danko R, Kmita A, Zymankowska-kumon S, Kubecki M, Gorny M (2014) Emission of polycyclic aromatic hydrocarbons (PAHs) and benzene, toluene, ethylbenzene and xylene (BTEX) from the furan moulding sands with addition of the reclaim. Metalurgija 53(4):451–454
17. Danko R, Gorny M, Holtzer M, Zymankowska-kumon S (2014) Effect of the quality of furan moulding sand on the skin layer of ductile iron castings. ISIJ Int 54(6):1288–1293
18. Kaminska J, Kmita A, Kolczyk J, Malatynska (2012) Strength parameters and a mechanical reclamation together with the management of its by-products. Metall Foundry Eng 38(2):171–178
19. Khandelwal H, Ravi B (2016) Effect of molding parameters on chemically bonded sand mold properties. J Manuf Processes 22:127–133
20. Ajibola OO, Oloruntoba DT, Adewuyi BO (2015) Effects of moulding sand permeability and pouring temperatures on properties of cast 6061 aluminium alloy. Int J Metals. Article ID 632021. http://dx.doi.org/10.1155/2015/632021
21. Bargaoui H, Azzouz F, Thibault D, Cailletau G (2017) Thermomechanical behavior of resin bonded foundry sand cores during casting. J Mater Process Technol 246:30–41
22. Johnston RE (1964) Statistical methods in foundry experiments. AFS Trans 72:13–24
23. Dabade UA, Bhedasgaonkar RC (2013) Casting defect analysis using design of experiments (DOE) and computer aided casting simulation technique. Procedia CIRP 7:616–621
24. Acharya SG, Vadher JA, Sheladiya M (2016) A furan no-bake binder system analysis for improved casting quality. Int J Metalcast 10(4):491–499
25. Parappagoudar MB, Pratihar DK, Datta GL (2007) Linear and non-linear statistical modelling of green sand mould system. Int J Cast Met Res 20(1):1–13
26. Surekha B Hanumantha, Rao D, Krishna G, Rao M, Vundavilli PR, Parappagoudar MB (2012) Modeling and analysis of resin bonded sand mould system using design of experiments and central composite design. J Manuf Sci Prod 12:31–50
27. Parappagoudar MB, Pratihar DK, Datta GL (2011) Modeling and analysis of sodium silicate-bonded moulding sand system using design of experiments and response surface methodology. J Manuf Sci Prod 11(1–3):1–14
28. Chate GR, Patel GCM, Deshpande AS, Parappagoudar MB (2017) Modeling and optimization of furan molding sand system using design of experiments and particle swarm optimization. Proc IMechE Part E: J Process Mech Eng. https://doi.org/10.1177/0954408917728636
29. Majumder A, Majumder A (2015) Application of standard deviation method integrated PSO approach in optimization of manufacturing process parameters. In Handbook of research on artificial intelligence techniques and algorithms, pp 536–563. IGI Global
30. Rao RV, Savsani VJ (2012). Mechanical design optimization using advanced optimization techniques. Springer, London. https://doi.org/10.1007/978-1-4471-2748-2

31. Rao RV (2010) Advanced modeling and optimization of manufacturing processes: international research and development. Springer Science & Business Media
32. Rao RV, Waghmare GG (2014) Complex constrained design optimisation using an elitist teaching-learning-based optimisation algorithm. Int J Metaheuristics 3(1):81–102
33. Patel GCM, Krishna P, Parappagoudar MB (2016) Modelling of squeeze casting process: conventional statistical regression analysis approach. Appl Math Model 40(15):6869–6888
34. Patel GCM, Krishna P, Parappagoudar MB, Vundavilli PR (2016) Multi-objective optimization of squeeze casting process using evolutionary algorithms. Int J Swarm Intell Res (IJSIR) 7(1):55–74
35. Rao RV, Kalyankar VD, Waghmare G (2014) Parameters optimization of selected casting processes using teaching–learning-based optimization algorithm. Appl Math Model 38(23):5592–5608
36. Venkata Rao R, Kalyankar VD (2012) Parameter optimization of machining processes using a new optimization algorithm. Mater Manuf Processes 27(9):978–985
37. Derringer G, Suich R (1980) Simultaneous optimization of several response variables. J Qual Technol 12(4):214–219
38. Maji K, Pratihar DK, Nath AK (2013) Experimental investigations and statistical analysis of pulsed laser bending of AISI 304 stainless steel sheet. Opt Laser Technol 49:18–27

Optimization of Electric Discharge Machining Based Processes

Roan Kirwin, Aakash Niraula, Chong Liu, Landon Kovach and Muhammad Jahan

Abstract The results of Electrical Discharge Machining (EDM) are characterized through many parameters. These include, material removal rate, surface finish, geometrical accuracy, tool wear, and kerf width. The three main types of EDM, wire, sinker, and micro EDM all have similar characteristics in relation to input parameters and their effects on the results. The typical EDM system is too complex to accurately model the effect of all the parameters together. Therefore, it is necessary to create an optimization algorithm to predict the results of specific input parameters. Various techniques such as Taguchi robust design, grey relational analysis, desirability, genetic algorithm, and neural network etc. have been used for optimization of EDM based processes. This chapter first briefly introduces all the aforementioned optimization processes and comprehensively discusses their implementation and effect for optimization of EDM based processes.

Keywords EDM · Wire-EDM · Optimization · Design-of-experiment · Surface roughness · Material removal rate · Fuzzy

1 Introduction

Electric Discharge Machining (EDM) is a non-traditional machining technique where the desired shape, size and geometry are obtained by thermoelectric erosion by electric sparks [1]. In general, machining is conducted by melting and vaporizing materials using electrical energy assisted by dielectric fluid. Dielectric fluid acts both as flushing source to carry melted and re-solidified craters, and as controlling agent for spark gap between electrode and workpiece. Workpiece and tool electrode act as cathode and anode, and maintain no contact throughout the machining process. For this reason, EDM is preferred when machining hard but electrically conductive materials as using conventional machining processes can be challenging.

R. Kirwin · A. Niraula · C. Liu · L. Kovach · M. Jahan (✉)
Department of Mechanical and Manufacturing Engineering, Miami University, Oxford, OH 45056, USA
e-mail: jahanmp@miamioh.edu

K. Gupta and M. K. Gupta (eds.), *Optimization of Manufacturing Processes*, Springer Series in Advanced Manufacturing, https://doi.org/10.1007/978-3-030-19638-7_2

EDM is categorized into mainly two types: Sinker EDM and wire EDM [2]. Sinker EDM is used to drill holes and cavities using an electrode of desired shape and profile. Sinker EDM uses profiled electrode to machine and is adopted for complex shapes as the electrode can be fabricated for various shapes. Wire EDM uses a wire to cut material, and is used to machine intricate shapes as the accuracy is extremely high. Micro-EDM is the variant of EDM where it is used at micro scale, along with micro level tools, axes resolution, and discharge energy [3]. The micro-EDM applies for both sinker and wire EDM, in addition to other micro-EDM varieties. Machining quality of EDM depends on several pre-built parameters. These parameters control the machining process, and can prominently change the output such as production time, surface roughness, over-cut, and dimensional accuracy in addition to desired material removal rate and electrode wear rate. Some of the common input parameters in EDM are [1–3]:

Gap voltage: Potential difference between anode and cathode for a cycle
Peak current: Maximum current that is used in machining per cycle
Pulse on time: Time duration for which current flows per cycle
Pulse off time: Time duration between two consecutive pulses
Duty cycle: Percentage of pulse on time over sum of pulse on time and pulse off time
Spark gap: Distance between electrode and the workpiece
Flushing pressure: Pressure at which the dielectric is dissipated

Given the impact of parameters and ever changing output demand, it can be difficult to find combination of variables that provides desired output. Numerous methods are used to approximate the favorable parameters. Statistical methods provide inexpensive and reliable means to test the effect of input variables to predict the desired outcome. In addition, some of the statistical methods like response surface method, and Taguchi method greatly aid in designing theoretical studies. This book chapter discusses the common statistical approaches in the beginning, and then transitions to provide overview of research works on optimization of EDM, based on various optimization techniques. In addition to studies on common wire and sinker EDM, descriptive review of research on optimization of smaller scale micro-EDM is also included in this chapter. Upon review of research works, the chapter ends by providing future avenues for subsequent research works on optimization of EDM parameters.

2 General Optimization Techniques

2.1 Single Variable Optimization

2.1.1 Taguchi Method

Taguchi method, sometimes called robust design method, is a statistical method that envisaged by Genichi Taguchi to optimize the quality of industrial goods, and recently was applied in the engineering, biotechnology and many other fields [4]. For the experimental design, Taguchi developed some well-structured guidelines. A set of arrays, named orthogonal arrays, is used in this method. These standard arrays could help the user find the minimum number of experiments that tell the full information of all the factors influencing the optimization results, i.e. performance parameters. Additionally, for each experiment, how the level combination of the input design variables is chosen, decides the crux of the orthogonal arrays method [5].

For the Taguchi method, the following steps need to be followed in order to design a good experiment.

1. Independent variable selection

Before the actual experiment is conducted, the researchers need to identify the key setting parameters that will influence the final performance parameters. For EDM process, different EDM machines need researchers to consider different setting of parameters. For example, when the wire EDM machine is used as the experimental setup, the parameters that researchers need to consider are pulse on time, pulse off time, peak current, gap voltage, servo voltage, servo feed rate, dielectric flow rate, wire speed or wire feed and wire tension. Depending on the potential research field, the researchers need to pick several or all of the setting parameters as the independent variables.

2. Number of level settings selection for each independent variable

Usually, the number of levels decided by getting the relation between the performance parameter and the independent variables. When the performance parameter has a linear relationship with the independent variables, 2 level settings are needed due to the fact that 2 points define a line. Similarly, when there is a quadratic relationship between the performance parameter and the independent variables, 3 level settings are needed. For all the nonlinear relationship, the number of level setting goes higher as the order of the relationship goes higher. For example, cubic relationship corresponds to a level setting of 4, quadratic relationship corresponds to a level setting of 5 and so on.

3. Orthogonal array selection

For orthogonal array, there are many standard arrays available, and each array corresponds to different number of independent design variable and level. For example, in Table 1, a typical L_9 orthogonal array layout is presented. In L_9 orthogonal array, the users aim to investigate the influence of 4 different independent variables with 3 set of (level) values followed. The array assumes that any of two factors are independent from each other. If there is any case, where the two factors have a strong interaction with each other, the orthogonal array would no longer be a suitable method for the experiment design.

In Table 1, there are 4 independent variables. And under each independent variable sections, there are 3 different level values. In order for user to get a better view of the influences of the independent variables, experiments need to be conducted

Table 1 Layout of typical L_9 orthogonal array [6]

$L_9(3^4)$ orthogonal array					
Experiment #	Independent variables				Performance parameter value
	Variable 1	Variable 2	Variable 3	Variable 4	
1	1	1	1	1	P1
2	1	2	2	2	P2
3	1	3	3	3	P3
4	2	1	2	3	P4
5	2	2	3	1	P5
6	2	3	1	2	P6
7	3	1	3	2	P7
S	3	2	1	3	P8
9	3	3	2	1	P9

under each combination. For example, in the first experiment, all the independent variables are set as level 1. After the experiment 1, the performance parameter value is recorded as p1. In the second experiment, the first independent variable is set as level 1, all the other independent variables are set as level 2 and the performance parameter value is recorded as p2. And so on for the rest of the experiments. Under these circumstances, a total of 9 experiments need to be conducted. And after all the experiments are conducted, 9 different performance parameter values are recorded. Once the results are achieved, the user needs to use optimized technique to find out optimum combination.

Depending on the numbers of the independent variables and the level of each independent variable, the orthogonal array has large number of variations. For two level designs, the Taguchi method has L4, L8, L12 and L16 orthogonal arrays; for three level designs, the Taguchi method has L9 and L27 orthogonal arrays; And for mixed level designs, the Taguchi method has L8, L16 and L18 orthogonal arrays. Additionally, for L16 orthogonal arrays, there are four different types of number of the independent variables and the levels combination. Once the orthogonal array type is decided, the overall structure of the whole experiment is outlined.

4. Experiment data and results analyzation

After all the experiments are conducted, the data are analyzed by the Taguchi signal to noise ratio (S/N) method. The signal to noise ratio is the ratio of mean value (signal) to the ratio of standard deviation (noise). The signal-to-noise ratio method is applied to determine the optimized settings based on the results. The S/N ratio has three different functions, larger is better, nominal is the best, and smaller is better. In the EDM machining parameters analysis, the S/N ratio of the smaller-the-better and the larger-the-better are the most useful characteristics and can be expressed as Eqs. (1) and (2) below [7].

$$(S/N)_S = -10\lg\frac{1}{n_k}\sum_{l=1}^{n_k} y_l^2 \tag{1}$$

$$(S/N)_L = -10\lg\frac{1}{n_k}\sum_{l=1}^{n_k} \frac{1}{y_l^2} \tag{2}$$

2.1.2 ANOVA

Analysis of variance (ANOVA) is a process by which a set of independent variables are analyzed for their influence on a single dependent variable. ANOVA is intended to demonstrate how strongly each independent variable is correlated with a change in the dependent variable [8]. At early stages of experimentation, ANOVA is used to narrow down which independent variables have the greatest effect and therefore should be studied. Also, later in the process, ANOVA is used to ensure that the results are statistically significant. A software package, such as R or SPSS [9], is typically used to perform ANOVA analysis because it is an extremely common method of statistical analysis. ANOVA produces both an F value and a percent of contribution for each independent variable. An F value references whether that parameter is significant at a specific confidence level. While a percentage contribution value measures the relative impact that a parameter has with respect to the rest of the parameters [10]. An example ANOVA table is included in Table 2.

2.1.3 Signal to Noise Ratio (S/N)

Signal-to-noise ratio (S/N) is used to optimize parameters once their individual impact on the independent variable is known. This ratio is commonly used in the

Table 2 ANOVA data table for surface roughness (Ra) [10]

Control factors	dof	Sum of squares	Mean squares	F-ratio	Percentage contribution
M	1	0.0012	0.0012	0.233	0.233
V_O	2	0.1267	0.0634	11.810	24.625
P_N	2	0.2683	0.1342	25.004	52.147
P_F	2	0.0172	0.0086	1.6051	3.343
V_S	2	0.0029	0.0015	0.2738	0.563
F_W	2	0.0438	0.0219	4.0786	8.531
T_W	2	0.0115	0.0058	1.074	2.235
P_D	2	0.032	0.0160	2.985	6.219
Error	2	0.0107	0.0054	–	2.122
Total	17	0.5145	–	–	100.00

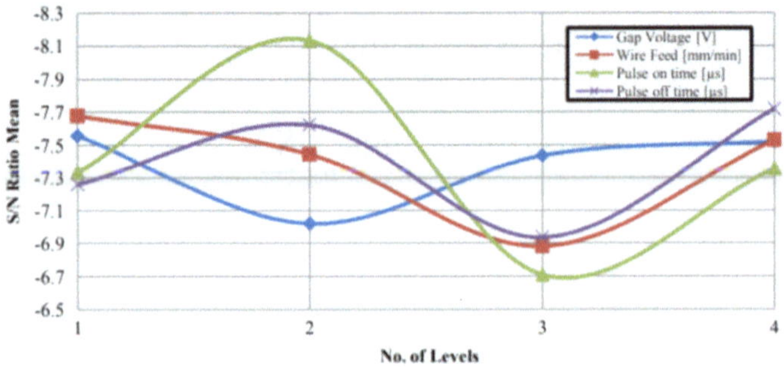

Fig. 1 Mean S/N ratio for effects of process parameters on surface roughness in EDM [11]

Taguchi method after ANOVA. S/N uses so-called control factors and noise factors. Control factors are defined as parameters that have a significant impact on the dependent variable while noise factors are defined as parameters that have no significant impact. S/N is a function of the output characteristic and the number of trials. This is demonstrated in Eq. (3) below.

$$\frac{S}{N_{LB}} = -10 \log\left(\frac{1}{r}\sum_{i=1}^{r} y_i^2\right) \tag{3}$$

where, r equal to the number of trials and y_i equal to the output characteristic. This specific equation favors lower value output characteristics. This would be used for dependent variables such as surface roughness or kerf width that are typically desired to be minimized. There are separate equations for characteristics that are to be maximized or be normal [11].

On a signal-to-noise ratio graph, the difference between these factors is very apparent because control factors will have a high degree of slope while noise factors have a small slope. Figure 1 shows an example of signal-to-noise ratio (S/N) chart. The optimal parameters are chosen by the highest mean S/N ratio for each set of parameters [12].

2.2 *Multiple Variable Optimization*

All of the aforementioned single dependent variable optimization methods are able to be adapted for multiple dependent variables. Typically, they are used as initial stepping stones to organize the data and then new set of organized data is analyzed again using other techniques such as multiple linear regression or neural networking. Other times, some normalization is applied so the dependent variables can be combined into one which can be optimized in the preceding ways.

2.2.1 Grey Relational Analysis (GRA)

The most common adaptation applied to the Taguchi method is grey relational analysis. Grey relational method is a data analysis method, which is used to measure the relation between the parameters. For two or more parameters where there is no information available in between, the situation is defined as black. For two or more parameters with perfect information, the situation is defined as white. In reality, neither of these situations happen in the experiment. So, for the situation in between the black and white, the relation is defined as grey. In the grey relational method, the degree of relation is defined as grey relational grade. For the situation of black relation, the grey relational grade is 0. For the situation of white relation, the grey relational grade is 1. And for the situation of grey relation, reality, the grey relational grade is between 0 and 1.

Typically, the grey relational method contains the following processing steps [13]:

- Normalize the experimental results of each response variable.
- Determine the grey relational coefficient for each response variable.
- Calculate the grey relational grade by the mean value of grey relational coefficient.
- Perform the response table and response graph for each level of the process parameters.
- Recognize the obvious and invisible variable factors and select the optimal level of the process parameter.
- Confirm test and verify the optimal levels of process parameter.

Mathematically finding the grey relational grades can be performed through a modification of the S/N formula. All data must be normalized by Eqs. (4) and (5) below before plugging it into the standard smaller the better S/N equation. With y_i = characteristic and k as the population size, the smaller the better equation (Eq. 4) is used, because the optimization direction is performed during the normalization step. The Grey relational coefficients that were fed into the modified S/N become the Gray relational grades. These grades are used as a measure of performance for multi response optimization [14].

$$x_i(k) = \frac{\max(y_i)k - (y_i)k}{\max(y_i)k - \min(y_i)k} \quad (4)$$

$$y_i(k) = \frac{(y_i)k - \min(y_i)k}{\max(y_i)k - \min(y_i)k} \quad (5)$$

2.2.2 Response Surface Methodology (RSM)

Response Surface Methodology (RSM) is a group of statistical methods that analyze the response of interest in terms of many different variables to optimize the response. RSM is an alternative of the Taguchi method. RSM was developed significantly earlier than the Taguchi method and requires more expertise but can outperform Taguchi in some situations. RSM allows for the analysis of the interaction of two independent variables simultaneously. This is typically realized by creating a second order response surface equation for each of the dependent variables. Occasionally, if the correlation is not strong enough for a simple second order polynomial equation to model, a log transformation could be required [15].

2.2.3 Multiple Regression Technique

The multiple regression technique is typically used to create models. After analysis by ANOVA or GRA or S/N, the parameters that are revealed to be significant can be related to the independent variables by multiple regression. All the independent variables must be normalized because they generally do not have the same units or ranges. This allows them to be directly related to the dependent variable through generally a polynomial relationship. These normalized values are typically called coded variables and vary from -1 to 1. A value of -1 corresponds to the minimum of this specific independent variable, while a value of 1 corresponds to the maximum value. Because insignificant variables are discarded, a simpler equation can be created similar to Eq. (6) below.

$$\hat{y} = \hat{\beta}_0 + \hat{\beta}_1 x_1 + \hat{\beta}_2 x_2 + \hat{\beta}_3 x_1 x_2 \tag{6}$$

where $\hat{y}$ represents the dependent variable, the β values represent the coefficients, and the statistically significant x values are the aforementioned coded variables. This case is for linear relationships that are not necessarily orthogonal. Higher β values indicate a higher influence of that coded variable on the dependent variable. Many papers rely solely on the P value to say that the results are statistically significant, but that is not necessarily true, and a method of residual should be performed to ensure statistically accurate result [16].

2.2.4 Desirability Function

Desirability function optimization is performed by first normalizing the response parameters into a desirability function d_i with the range $0 \geq d_i \geq 1$. When the response parameter is at the desired target then $d_i = 1$ and if it is outside a specified range $d_i = 0$. Each desirability function is then assigned a weight, r_i. The weight of each of the desirability functions changes its polynomic order. All of the desired ability functions are then combined using a weighted geometric mean. This is demonstrated in Eq. (7) below.

$$D = \left(d_1^{i_1} d_2^{i_2}\right)^{1/(i_1+i_2)} \tag{7}$$

The majority of statistical software, such as Minitab®, use a reduced gradient algorithm with multiple starting points to maximize the overall desirability function.

However, this is not necessarily the true optimal solution if the boundary is non-convex. A local maximum may be achieved, so some manual tweaking is necessary in this method. This can be done by changing the weighting of the individual desirability functions [15].

2.2.5 Pareto Optimization

Pareto optimization is based on a concept called non-inferiority. Non-inferiority is defined as being reached when improving one response absolutely requires the decrease in quality of another response variable. As this is a multi-objective optimization, there are more than one "optimal" solution and this manifests itself as an optimal front. Imagine a scatter plot with surface roughness on one axis and 1/MRR on the other. Graphed on this plot are a series of results from parametric testing. The data will be distributed in a cloud. If each data point that were on the bottom left side of the cloud were connected, this would form an optimal front. This process can severely reduce the problem size of an optimization problem. It can be applied to not only individual response variables but also normalized and combined variables allowing many more response dimensions [15].

2.2.6 Genetic Optimization Algorithms

Genetic algorithms (GA) perform optimizations through an evolutionary approach. In general, they work by randomly seeding an initial population of random weights for each response parameter. The best individual is chosen and a new population is generated by randomly applying a certain amount of mutation to the individual. This process is repeated either a set number of times or until the rate of convergence reaches a set threshold [7]. A common application of genetic optimization is called the Non-Dominated Sorting Genetic Algorithm (NSGA). This adapts Pareto Optimization to work with the genetic algorithm to ensure only non-inferior solutions are found. The non-inferior front that is found in Pareto optimization is also called the non-dominated solution set.

The process of a non-dominated genetic sorting algorithm is as follows:

1. A set of chromosomes are generated within the solution space that have all the input variables on each chromosome. This is the initial population.
2. These chromosomes are sorted based on the following rules, if satisfied they are marked dominated else they are non-dominated.
    a. Response A of Chromosome A < Response A of Chromosome B
    b. Response B of Chromosome A < Response B of Chromosome B
3. All non-dominated solutions are ranked 1.
4. The sorting is repeated with already ranked chromosomes removed and new non-dominated chromosomes are ranked 2.

5. This process is repeated until all chromosomes in the population are ranked.
6. A dummy fitness value is assigned to the rank 1 chromosomes.
7. The normalized Euclidean distance is calculated between each rank 1 chromosome and the rest of the rank 1 subpopulation.
8. The Sharing function values are calculated for each of this subpopulation which are then used to calculate the niche count. Sharing functions are based on the largest distance allowed between chromosomes in a niche group. The niche count is used to estimate the clustering close to a specific chromosome.
9. The dummy fitness values are then divided by the niche count to find the shared fitness value of the rank 1 subpopulation.
10. This process is repeated until all ranks are assigned a shared fitness value.
11. The fitness values are used to choose which chromosomes to mate and mutate to form the next generation population.
12. This entire process is repeated on the next population.

Non-dominated Sorting Genetic Algorithm (NSGA) will generate varying results depending on the parameters chosen such as population size and mutation rate. Higher mutation rate will typically lead to a faster convergence but a less accurate one while a larger population size will lead to a higher convergence rate at the expense of computational power [17].

2.2.7 Neural Networking

Neural networking works similarly to genetic algorithm optimization in that it is an iterative process. Neural networking simulates the learning process of neurons, connections that lead to advantageous results are reinforced while connections that are useless or have a negative impact fade in strength. These neurons are made up of levels. The first set of neurons are the input neurons that receive the input parameters which, in the case of WEDM, are things such as pulse on time, wire tension, and open voltage. Each of the input neurons is then linked to the first 'hidden' layer. There can be one or more hidden layers depending on how the input parameters are related to each other and the guessed order of the input output relation. More complex relationships between the input variables and the response variables require more hidden layers to model. The drawback of more hidden layers is a longer convergence time and a less accurate result. Typically, the lowest complexity model that gives an effective result will be the best model. Using overly complex models or overtraining a model can lead to overfitting, which looks like a very effective model for the data already collected but is a very poor predictor of new data. Each of the connections between the two layers is assigned a weight and these weights are adapted using a system similar to the genetic algorithm optimization discussed above [8]. Figure 2 shows the block diagram listing the steps for optimization using back propagation neural networking [18].

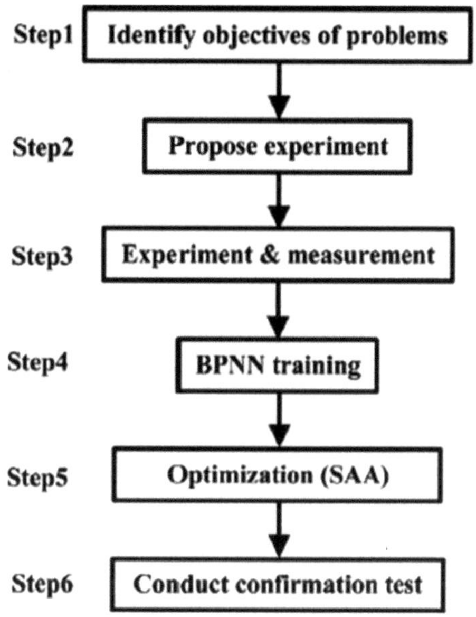

Fig. 2 Flow chart showing the steps of back propagation neural networking (BPNN) for optimization [18]

2.2.8 Simulated Annealing

Simulated annealing is applied to neural networks to ensure a speedy convergence while allowing for a more accurate result than a typical high entropy neural network training scheme. Its namesake, metal annealing, is the slow reduction of heat from a chaotic state to a cooler more rigid state. The slow cooling rate allows crystals to form and the final state to be highly ordered. With a very high cooling rate crystals do not have time to grow and the final metal is amorphic with the metal stuck in a high energy state. Simulated annealing applies these ideas to a neural net. As the system is trained the entropy allowed between generations is decreased so large changes can be made early and the model can be generally optimized. As iterations pass the system entropy is 'cooled' which simply means that the size of changes between generations, or the mutation rate, is decreased. This allows the system to make smaller changes, as it approaches what is theoretically the optimal, or the most ordered crystalline, solution [19]. In many cases, the simulated annealing is integrated with the neural network to perform the optimization. Figure 3 shows the flow chart of an integrated neural network and simulated annealing.

3 Optimization Techniques Used in Wire EDM

Most wire electro-discharge machining (WEDM) optimizations seek to increase material removal rate (MRR) or decrease the kerf width while decreasing surface

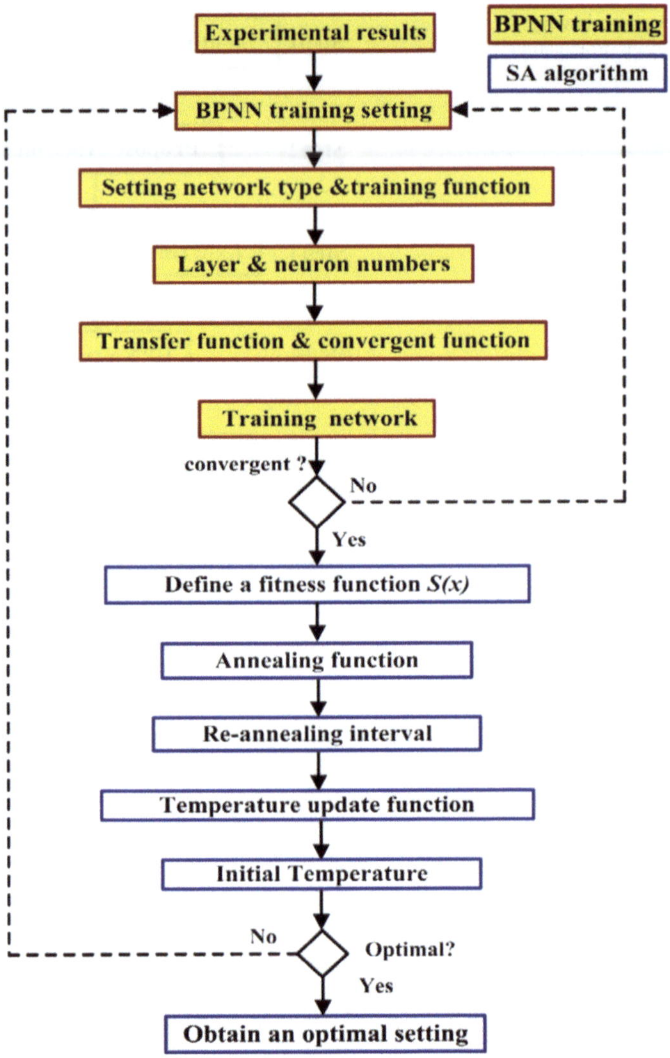

Fig. 3 Integrated neural network and simulated annealing steps for optimization [18]

roughness. The requirement of optimizing two variables simultaneously greatly complicates the process. The Taguchi method was developed in order to optimize a single dependent variable [20]. In order to optimize more than one variable another statistical technique is needed. Common analyses include Grey relational method, neural networking, response surface methodology, the desirability function, and Pareto optimality. While the majority of optimizations in this space are optimizing for at least two characteristics, sometimes it is desired that a single variable be optimized. Such cases are simpler to optimize using a standard Taguchi method or another method such as the finite element method.

3.1 Taguchi Method, ANOVA, S/N Ratio

As the Taguchi method is more oriented towards experimental design, all of the single variable optimization methods used Taguchi to create the initial parametric layout of the experiments. Goswami and Kumar [21] optimized the wire-EDM process for machining of Nimonic alloy using Taguchi approach along with utility concept. Taguchi's robust methodology was used for design of experiments and multi-response optimization method was used to study the effect of pulse on time, pulse off time, and peak current on the material removal rate (MRR), surface roughness (SR) and surface topography. The optimized process conditions were identified for MRR and SR using both single- and multi-response optimization techniques. In addition, the effect of parameters on microstructure and recast layer under the machined surface was studied. It was found that multi-response optimization with utility concept provided optimized parameters that could be used to improve the wire EDM performance.

Ramakrishnan and Karunamoorthy [22] optimized the wire-EDM operation for machining tool steel using multi-response optimization methods with Taguchi's robust design approach. The experiments were planned using Taguchi's L16 orthogonal array to study the effect of wire-EDM operating parameters on MRR, SR, and wire wear ratio. ANOVA was applied to study the level of importance of each machining parameter on the wire EDM performance. It was found based on ANOVA that pulse on time and ignition current intensity had more significant influence on wire EDM performance than other operating parameters. It was reported that the proposed optimization method could successfully predict the wire-EDM performance and could be applied to improve the machining performance.

Manna and Bhattacharyya [12] investigated the effect of operating parameters experimentally and optimized the process using Taguchi's optimization technique during wire-EDM of aluminum-reinforced silicon carbide metal matrix composite (Al/SiC-MMC). A Taguchi L18 orthogonal array was used to design the experiments and identify S/N ratio, and then ANOVA and F-test were used to identify the significant parameters affecting the wire EDM performance. Mathematical models were developed using the Gauss elimination method and compared with the experimental results. The model was found to successfully predict the optimum conditions suggested by the experiments.

Mahapatra and Patnaik [23] optimized the wire-EDM process for machining D2 tool steel using the Taguchi method. The significant parameters influencing the wire-EDM process were identified and their effect on wire EDM performance of D2 tool steel was studied. The relationship between the input machining parameters and output performance parameters were established using model developed by non-linear regression analysis. It was demonstrated that the optimization technique enabled adjustment of machining parameters to obtain higher material removal rate while minimizing the surface roughness during wire-EDM of D2 steel.

Tilekar et al. [24] optimized the process parameters for wire EDM of aluminum and mild steel using Taguchi method for obtaining minimum surface roughness and kerf width. Single objective Taguchi method was used for optimization of process parameters, while ANOVA was used to test statistical significance of each process parameter on influencing the wire EDM performance. The ANOVA results show that spark on time and wire feed rate has significant effect on kerf width, whereas spark on time and input current influences the surface roughness more significantly.

ANOVA was also used in almost all papers as an initial analysis of the impact of specific parameters on response variables. Kanlayasiri and Boonmung [16] used design of experiments, ANOVA, and regression model to investigate the impact of machining parameters on surface roughness of a new die steel, DC53. The ANOVA was used to identify the parameters that influenced the wire EDM process most significantly. It was determined that pulse on time and peak current had the largest impact on the surface roughness. The mathematical model was developed by multiple regression method, and was validated by experimental data with the prediction error less than 7%.

Signal-to-noise ratio (S/N) is the most common way to optimize parameters after performing ANOVA. Manna and Bhattacharyya [12] attempted to optimize MRR, surface roughness, gap current, or spark gap in WEDM of Al/SiC-MMC composite using S/N. As S/N optimization does not create a true model, it can only optimize the tested parameters without any interpolation between values, the returned optimization is in terms of the original experimental values. This specific study was able to optimize for each of the desired response parameters. Figure 4 shows the comparison of experimental results with the predicted results obtained from developed mathematical models [12]. It can be seen that the proposed mathematical model can predict the outcome of the experiments quite successfully during wire EDM of Al/SiC-MMC.

Ikram et al. [10] sought to optimize effect of eight operating parameters on MRR, SR, and kerf width during WEDM of D2 tool steel. This was performed using Taguchi's design of experiment (DOE), ANOVA, and S/N. The S/N returned an optimal setting for each of the eight parameters for MRR or surface roughness. It was deduced that higher MRR required a higher pulse on time and open voltage, while a lower surface roughness required the opposite.

3.2 Grey Relational Analysis

Rajyalakshimi and Venkata Ramaiah [25] strove to optimize MRR, surface roughness, and spark gap in WEDM of Inconel 825. Optimization using Grey relational analysis on nine input parameters and three output parameters were able to improve MRR from 119 to 126 mm^3/min, spark gap from 15 to 13 μm, and surface roughness from 1.68 to 1.44 μm.

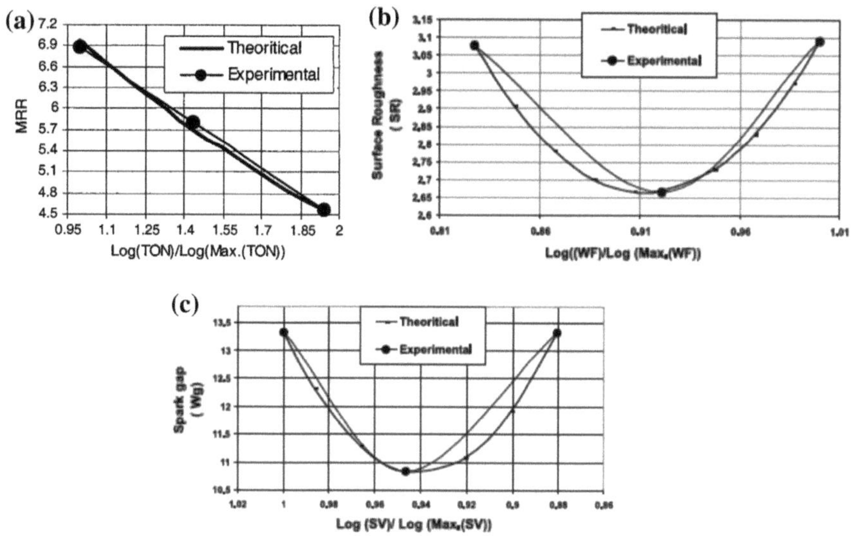

Fig. 4 Comparison of experimental results with developed mathematical models for **a** MRR, **b** surface roughness, and **c** spark gap [12]

Bobbili et al. [14] attempted to optimize WEDM machining of ballistic grade aluminum alloy with four input parameters and three response parameters. Using grey relational analysis, three of the input parameters were found to have a significant impact and were optimized. Significant improvements in MRR, surface roughness, and gap current were observed with a 6% error.

Durairaj et al. [11] applied grey relational theory and Taguchi optimization technique to optimize the cutting parameters during wire EDM of SS304 steel. The goal of the optimization was to identify optimum parameters for minimum kerf width and best surface finish separately and simultaneously. Taguchi's L16 orthogonal array was used for designing experiments. It was found that the Taguchi optimization and grey relational theory provided two different settings of optimized parameters. In Taguchi optimization technique, the parameters combination for minimum surface roughness were 40 V gap voltage, 2 mm/min wire feed, 6 μs pulse on time, 10 μs pulse off time, and for minimum kerf width the combination was 50 V gap voltage, 2 mm/min wire feed, 4 μs pulse on time, 6 μs pulse off time. According to grey relational analysis, the optimized parameters setting to get both the minimum surface roughness and the nominal kerf width were 50 V gap voltage, 2 mm/min wire feed, 4 μs pulse on time and 4 μs pulse off time.

Huang and Liao [5] carried out optimization of wire EDM process for machining SKD 11 tool steel using a combination of Taguchi method, Grey relational analysis and S/N ratio. First the experiments were design using Taguchi's L18 orthogonal arrays, then the parameters were optimized using Grey relational analysis and S/N ratio for obtaining the maximum material removal rate and minimum surface rough-

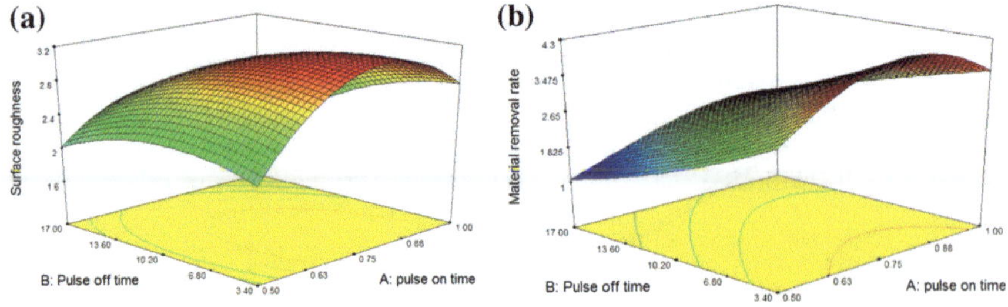

Fig. 5 Surface plots showing interactions of pulse on time and pulse off time with **a** surface roughness, and **b** material removal rate [26]

ness in wire EDM of SKD 11. The statistical analysis, including S/N ratio, ANOVA, and F-test, was carried out to find out the significant parameters for wire EDM.

3.3 Desirability Function and Pareto Optimization

Raj and Senthilvelan [26] optimized the wire EDM machining conditions using desirability function with a goal to improve the surface finish and material removal rate. They used Box-Benkhen approach to design the experiments, and desirability function to empirically model and optimize the process parameters for wire EDM of Ti-6Al-4V alloy. It was found that pulse duration and pulse interval were two important parameters influencing the surface roughness most, whereas pulse interval had the most influence on material removal rate. Higher pulse interval decreased the material removal rate significantly. Figure 5 shows the surface plot representing the influence of pulse on time and pulse off time on the surface roughness and material removal rate during wire EDM of Ti-6Al-4V, as obtained by desirability function optimization [26].

Sarkar et al. [27] sought to model and optimize WEDM of γ-titanium aluminide alloy during trim cutting. Trim cutting was used to increase the surface finish after a roughing operation so the desired response parameters would be below a chosen surface roughness while at the highest machining speed possible. First, the desirability function was used in Minitab and then a Pareto optimization was performed. Both of the techniques were effective at modeling the relationships but the desirability approach required a lot more manual fiddling and tuning than the Pareto optimization approach. Figure 6a shows that multiple optimal points were identified in the Pareto optimization, and when plotted (Fig. 6b), they showed a trend of increasing the surface roughness with the increase of cutting speed [28].

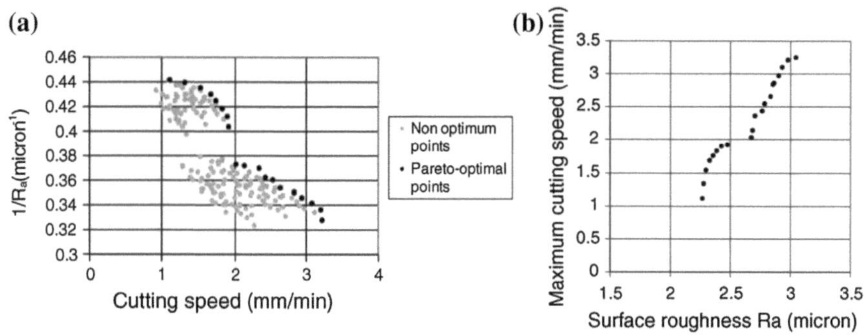

Fig. 6 **a** Multiple optimal solutions for surface roughness based on cutting speed, as suggested by the Pareto-optimal solution, **b** Plot of optimal points in the form of maximum cutting speed versus surface roughness [28]

3.4 Genetic Optimization Algorithms

Mahapatra and Patnaik [29] used the Taguchi method along with ANOVA and S/N as an initial optimization and for data gathering. They then used this data to train a genetic algorithm to predict MRR and surface finish. A model of MRR and surface finish was created with multi regression and were then combined and ran through a genetic algorithm to optimize the input variable to achieve the desired response parameters. The optimal settings for MRR and surface finish agreed well with the confirmation experiments with 4 and 1.5% errors respectively.

Kumar and Agarwal [30] used a multi-objective genetic algorithm to find optimal solutions for MRR and surface finish. The experiments were designed based on the Taguchi's design of experiments to study the effect of operating parameters on wire EDM performance of high speed steel (SKH9). The mathematical model has been developed by non-linear regression analysis. They were able to use the Pareto optimization technique along with non-dominated sorting genetic algorithm (NSGA) to further increase the efficiency of the solutions. Figure 7 shows the proposed multi-objective optimization methodology used in the study [30]. They were able to develop a table with 50 different optimal solutions that would then allow the operator to choose the set that best matches their requirements. Out of 50 optimal solutions, the best parametric combination providing highest MRR while maintaining the minimum surface finish requirement is pulse peak current of 30 A, pulse duration of 37 μs, pulse off time of 50 μs, wire feed of 7 m/min, wire tension of 1260 g, flushing pressure of 2.1 kg/cm^2.

Kuriakose and Shunmugam [17] optimized the wire-EDM process for machining titanium alloy using Non-Dominated Sorting Genetic Algorithm (NSGA) method. A multiple regression method was used to study the effect of operating parameters on the wire EDM performance, while NSGA method was used for optimizing the process. It was found that a number of optimized set of solution could be obtained using the

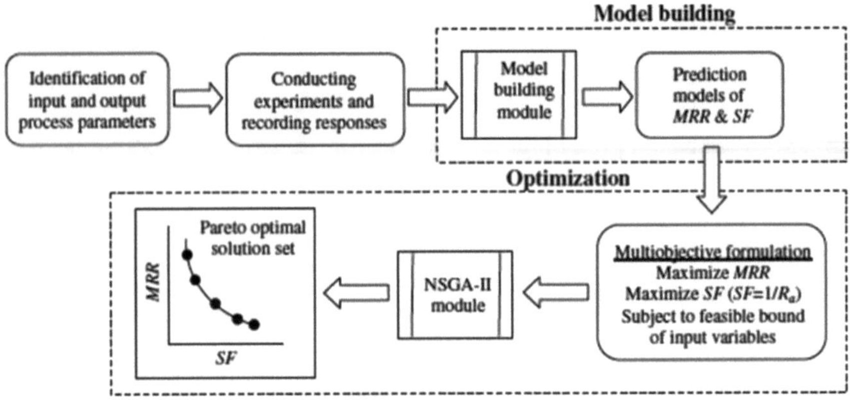

Fig. 7 Proposed multi-objective optimization methodology used in the wire EDM of high speed steel SKH9 [30]

Pareto optimization method. They proposed a table with 36 optimal combination of parameters.

Rao et al. [31] carried out experimental investigation and parametric optimization of wire EDM process for machining Al2014T6 aluminum alloy. The Taguchi method was used to design the experiments and a hybrid genetic algorithm with linear regression model was used for optimization of machining parameters for improved surface finish and material removal rate. It was found that the developed model could successfully predict the machining performance. It was also found from the genetic optimization that the cutting efficiency was important for generation of good quality surface finish. It was also reported that the recast layer in aluminum alloy was comparatively higher than those reported in wire EDM of heavier materials and other lighter material like titanium alloy.

3.5 *Response Surface Methodology, Neural Networking and Simulated Annealing*

Response surface methodology (RSM) is a comparatively old method of optimization, so it is often used to compare to neural networking optimizations. Speeding and Wang [32] used both RSM and a back-propagation neural network to optimize MRR, surface roughness, and surface waviness. Both models were found to achieve an acceptable level of accuracy with neural networking achieving a slightly higher accuracy. The prediction errors of the neural network model are generally lower than the RSM model.

Yang et al. [33] investigated both RSM and neural networking optimization, but with a more advanced simulated annealing neural networking system for optimizing the wire EDM process. Taguchi L18 orthogonal array was used for designing the

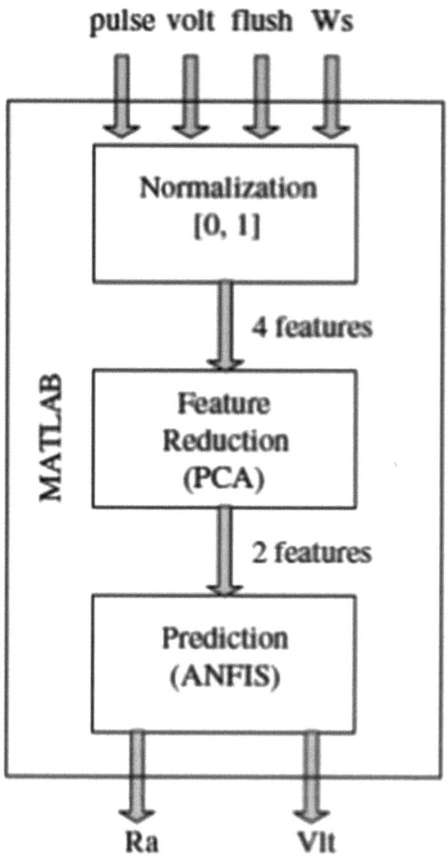

Fig. 8 Block diagram showing different stages of the proposed adaptive neuro-fuzzy interface system (ANFIS) model [34]

experiments before optimizing with RSM and BPNN. The neural network provided a more accurate predictive model than RSM did, but the neural network took more experience and training to implement.

Çaydas et al. [34] developed an adaptive neuro-fuzzy interface system (ANFIS) model to predict the white layer thickness and surface roughness as a function of machining parameters during wire EDM of AISI D5 tool steel. The dual approaches allowed the solution to converge faster and to improve the accuracy of the model. The model and verification experiments were closely correlated within a small margin of error. Figure 8 shows the steps of ANFIS optimization model used in this study. The comparison of the predicted results of surface roughness and white layer thickness with those obtained from experimentation is shown in Fig. 9. It can be seen from the graphs that the proposed model was able to predict the surface roughness and white layer thickness with minimal error.

Simulated annealing, a modification of standard neural networking, is commonly used to allow a fast convergence while still getting an accurate model. Chen et al. [18] developed a model of WEDM of pure tungsten using the simulated annealing assisted

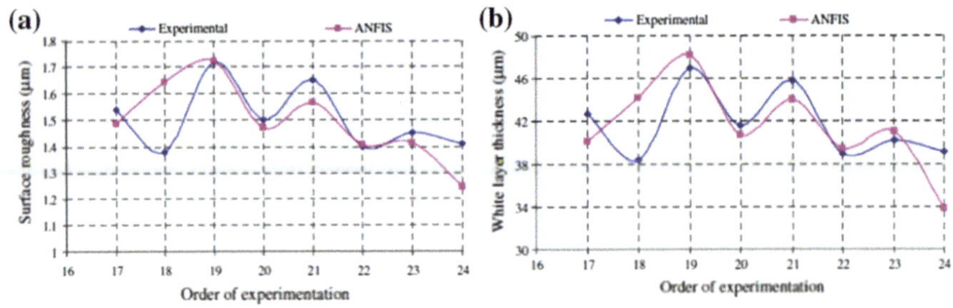

Fig. 9 Comparison of predicted results from ANFIS model with experimental results, **a** surface roughness, and **b** white layer thickness [34]

neural networking approach. In their study, a back-propagation neural network with simulated annealing method was proposed to optimize the parameters for improved surface finish and cutting velocity. This model was extraordinarily accurate for both the average roughness and maximum roughness, but the cutting velocity model was almost six orders of magnitude less accurate. Figure 10 shows the configuration of the proposed back propagation neural network model for wire EDM optimization [18].

Tarng et al. [35], quite a while ago in 1995, developed a simpler neural network to model the WEDM of SUS-304 stainless steel. This paper was written when neural networking was new. As a result, it was more of a proof of concept, but they proved that simulated annealing could indeed be used to increase the efficiency of Wire EDM.

Spedding and Wang [36] modeled the surface generated in wire EDM process using artificial neural network and time series techniques. The experiments were designed using central composite design (CCD). The feed-forward BPNN was used to develop the model and optimize the process. The optimal combination of process parameters for improved surface finish was identified and the wire EDM surface profile were evaluated by predicting the periodic component of the surface.

4 Optimization Techniques Used in Die Sinker EDM

4.1 Response Surface Method

Response surface method is a statistical process to estimate response variables using input parameters. It is useful in quantifying the connection between input and output parameters and is cost effective. Using response surface method, Balasubramanian et al. [37] conducted a study to assess the effects of peak current, pulse on time, dielectric pressure and tool diameter on material removal rate (MRR), surface roughness (SR), and tool wear rate (TWR). Cast and sintered copper electrodes were used to

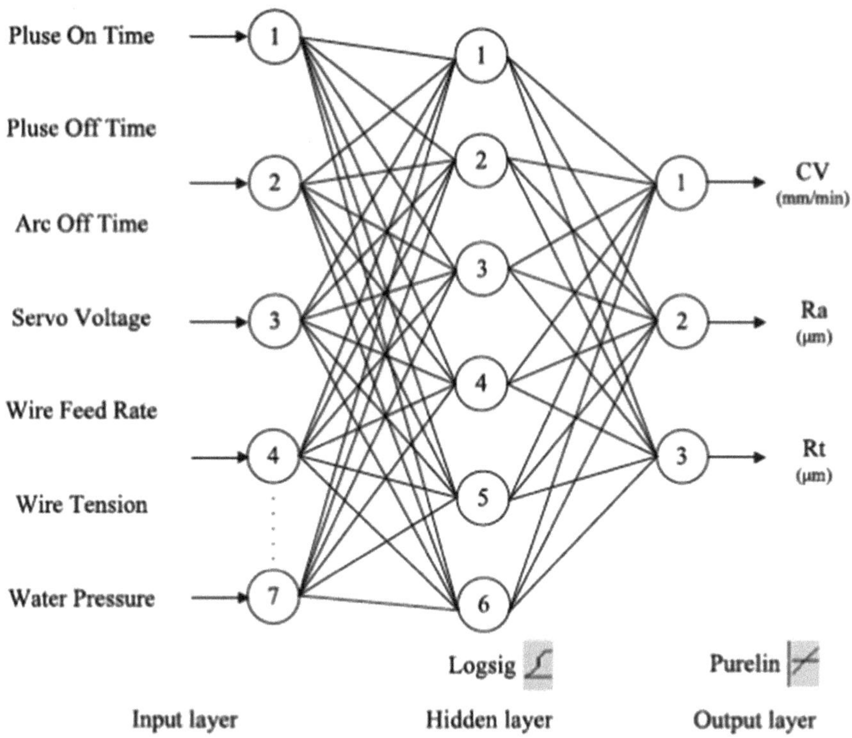

Fig. 10 The configuration of the proposed back propagation neural network (BPNN) model used for optimizing the wire EDM process [18]

machine EN8 and D3 steels. It was found that increasing the pulse on time increased surface roughness with cast copper electrode and increasing peak current increased MRR and TWR on EN-8 and D3. Tool diameter had a directly proportional relationship with MRR, TWR and SR. Both EN-8 and D3 showed higher MRR and lower TWR with cast electrode in comparison to sintered copper electrode. Surface roughness mean values were found to be lower with sintered electrode. High peak current and pulse on time with larger electrode diameter provided high MRR and SR, but low TWR was obtained with sintered copper electrode on EN8. The cast copper electrode provided better MRR, TWR and SR on D3 with low pulse on time and dielectric pressure.

Mishra et al. [38] focused on investigating the effect of hardness on MRR and TWR during die-sinking EDM of EN31 steel using copper electrode. During the experiment, four parameters, pulse on time, pulse off time, peak current and gap voltage, were varied under response surface method to identify changes in response variables. It was found that more than 70% out of 30 experiments had increased machining time for hardened workpiece including increase in tool wear rate as displayed in Fig. 11.

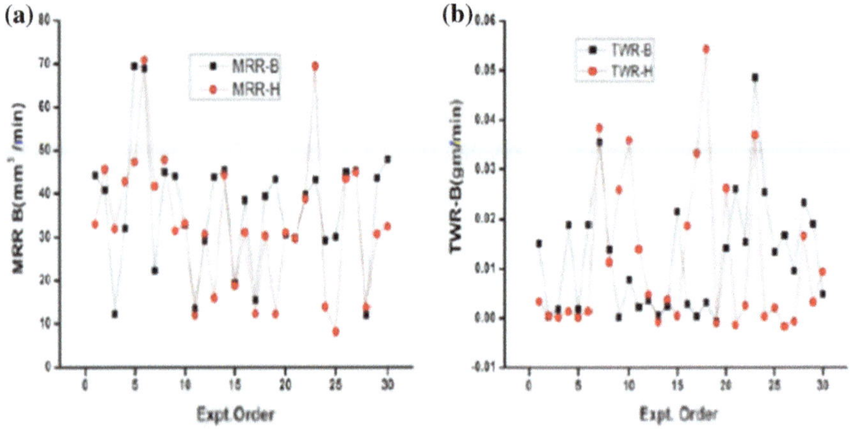

Fig. 11 Comparison of **a** MRR and **b** TWR of base alloy and hardened alloy after die sinking EDM using copper electrode [38]

Payaghan et al. [39] utilized response surface method using current, voltage, pulse on time and duty factor as input parameters during EDM of copper tungsten matrix composite. Copper electrode with Rush lick-30 dielectric fluid was used during the die-sinking operation. It was observed that increasing discharge current and pulse on time increased surface roughness. Overheating and molten metal in addition to discharge column expansion were suggested as the reasons for increase in surface roughness. The optimal machining condition resulted in surface roughness of 3.62 μm. All the results presented had over 95% confidence interval.

Leao et al. [40] explored the effect of different electrodes and dielectric fluids on the electrode life during fast hole EDM drilling of nickel based workpiece. Deionized water and a mixture of water, alcohol and salts were two di-electric fluids, and copper and brass were two electrodes used in this study. With the help of Pareto chart, it was found that drilling time was mostly affected by duty cycle and peak current, along with the interaction of dielectric with duty cycle and peak current. Electrode wear was greatly influenced by the dielectric and duty cycle. Heat generation due to high peak current resulted in low drilling time but high electrode wear. Brass electrode and water mixture dielectric produced the best results with only 47% of electrode wear. Deionized water had higher breakthrough time due to electrode tapering. Overall, the impact of dielectric fluid in optimization of fast hole drilling was justified in the study.

4.2 *Taguchi Method*

One of the frequent challenges during EDM machining is the tool wear. To provide accurate machining and smoother surface, it is important to gain minimal tool wear by optimizing the process parameters. Urade and Deshpande [41] investigated TWR during die-sinking EDM of EN 31 alloy steel using copper electrode. Parameters such as pulse current, pulse duration, and gap voltage were varied and studied for L16 orthogonal array. It was found that 3 A discharge current with 110 μs pulse duration and 130 V provided optimal quality in regards to TWR. Pulse current was noted to be the most significant factor affecting the response variable.

Lee et al. [42] studied the effect on machining parameters on SR and TWR using Taguchi method while also incorporating utility concept. It was found that current and duty cycle had notable effect on both SR and TWR, during die-sinking EDM of EN31 steel using copper electrode. Interactions between voltage and pulse on time and duty cycle were significant in addition to spark gap and pulse on time. The optimal parameters for die-sinking EDM of EN31 steel using copper electrode were found to be current of 6 A, duty cycle of 9 and pulse on time of 200 μs. Further experiments were conducted to analyze the response variable results with the optimal parameters. Utility value was found to be 8.289, and average surface roughness and average tool wear rate at the optimal setting were found to be 5.102 μm and 0.0092 g/min respectively.

One important characteristics of sinker EDM is flushing method, as it is critical to clean eroded particles from the machining zone. Malhotra et al. [43] utilized slide flushing to study the effect of flushing along with other machining parameters on the surface roughness of EN-31 die steel using die-sinker EDM. Copper electrode was used during the study. L27 orthogonal array was used using input parameters such as current, voltage, pulse on time, duty cycle, spark gap and flushing pressure to study the effects on surface roughness. It was found that current in addition to pulse on time had greater influence on the surface roughness. Interaction between pulse on time and spark gap was also significant. The optimal parameters to produce minimum surface roughness were found to be a combination of 6 A current, 100 μs pulse on time, 0.5 mm spark gap, 0.6 kg_f/cm^2 flushing pressure, and 35 V voltage.

Rath [44] conducted L27 orthogonal array based experiments on EN19 alloy steel using copper electrode on die-sinker EDM. The parameters used were pulse on time, current, and pulse duty factor, and the performance parameters were MRR, TWR, SR, and overcut (OC). It was found that pulse on time was the most dominant parameter that changed MRR and TWR results. MRR was seen to increase with increase in pulse on time to a certain extent, and reduced as the energy transfer was difficult with formation of plasma. OC was seen to increase with increase in pulse on time and open circuit current due to extreme melting and evaporation. Grey Relational Analysis was also utilized to obtain optimal parameters, and they were found to be open circuit current of 30 A, pulse on-time of 3000 μs, and pulse duty factor of 12.

According to the study by Prasad et al. [45] on AISI P20 tool steel using die-sinker EDM with copper electrode, MRR increased with increase in pulse current. The analysis of the input parameters, i.e. current, voltage, pulse on time, and pulse off time, were performed using L9 orthogonal array of Taguchi method. Higher pulse on time was attributed to increased MRR due to thermal power improvement, and higher current led to excessive material erosion. It was noted that increase in both voltage and current increased MRR, while TWR was decreased with voltage and increased with current.

Belgassim et al. [46] evaluated the influence of pulse current, pulse on time, pulse off time, and gap voltage on EDM of AISI D3 steel using L9 orthogonal array of Taguchi method. Primary variables of output interests were surface roughness and over-cut. Die-sinker EDM with brass electrode and kerosene dielectric were utilized during machining. Using S/N ratio, pulse current was found to influence the surface roughness the highest. Optimal parameters for best surface finish were determined to be 26 A current, 50 μs pulse on time, 200 μs pulse off time, and 45 V voltage. For over-cut pulse on time was the most influential factor as well.

Shrivastava et al. [47] investigated SR, MRR, and TWR on one of the widely used AISI 202 stainless steel using die-sinker EDM with copper electrode. Taguchi's L9 orthogonal array of four factors and three levels input variables such as pulse on time, pulse off time, peak current and servo voltage were applied. It was seen that pulse on time and servo voltage were the essential factors that affected surface roughness.

Panda et al. [48] used die sinking EDM with a copper electrode on 6061 aluminum alloy. Three input parameters, i.e. duty cycle, pulse on time, and current, were used in Taguchi method to study their effect on MRR. It was found that increase in pulse on time and duty cycle increased MRR. Optimal process parameters were found to be 100 μs pulse on time, 10 duty cycle and 30 A current. Overall, current had the largest impact on MRR.

Instead of using common method of submerged dielectric for machining in die-sinking EDM, Shahril et al. [49] tested the alternative by spraying dielectric fluid during machining. Taguchi method was used to study effect of discharge current, pulse on time and pulse off time on the MRR. Graphite electrode was used to machine tool steel workpiece in the presence of kerosene dielectric fluid. It was found that spray method provided lower machining time but higher surface roughness in comparison to submerged method. The MRR was observed to be higher as pulse on time increased, and found to be lower with increase in pulse off time.

Sangeetha et al. [50] explored die-sinking EDM on different aluminum base metal matrix composites under change in parameters such as current, pulse on time, pulse off time, and tool lifting time. L27 orthogonal array was utilized for designing the experiments with a copper electrode. It was found that reinforcement material for composite, current, and tool lifting time were highly influential in increasing MRR, and reducing SR, cost, and TWR.

Dubey et al. [51] investigated a different composite made of Al and Al_2O_3 to find the optimal parameters for machining using die sinker EDM with copper electrode. The input variables that were analyzed using desirability function and Taguchi method included pulse duration, discharge current, and duty cycle. The effect of operating parameters on the MRR, SR and TWR were also studied. The study confirmed the effectiveness of using desirability function with Taguchi method in a production system to maintain quality control.

Gaikwad et al. [52] studied the effect of input variables, i.e. gap current, pulse on time, pulse off time, workpiece electrical conductivity, and electrode conductivity during EDM of NiTi alloy using copper electrode. The experiments were conducted

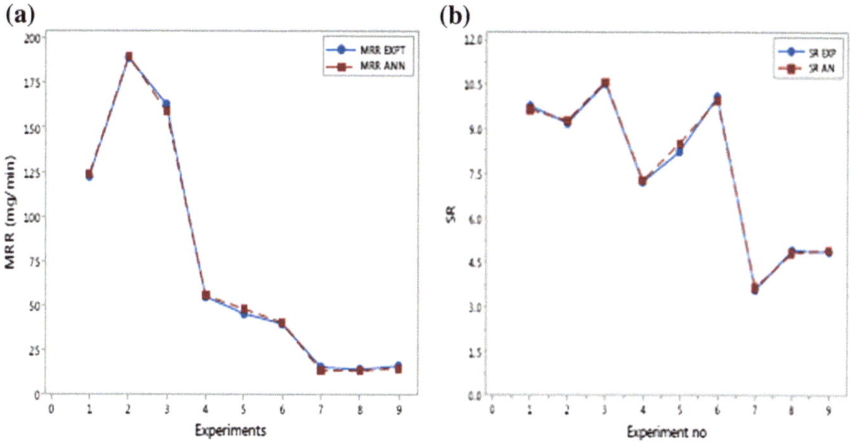

Fig. 12 Comparison of predicted ANN model with experimental results for **a** MRR (tested) and **b** SR (tested) [54]

based on Taguchi's L36 orthogonal array. It was found that gap current, work electrical conductivity and pulse on time had significant effect on MRR. Optimal parameters for obtaining MRR of 7.0806 mm^3/min were found to be 4219 S/m electrical conductivity, 16 A gap current, and 38 μs pulse on time.

Daud [53] carried out experiments with changes in current rates to see the effect on surface of SKD 11 using die-sinker EDM with hollow copper electrode. In the study, pits were machined, and the results of variable current rates were analyzed using Taguchi method. It was found that there was notable relationship between current rate and radius of workpiece. Higher current magnitude led to anomalies in the machined radius due to increase in recast later.

4.3 Artificial Neural Network (ANN)

Chandramouli and Eswaraiah [54] studied the influence of artificial neural network (ANN) on determining machining standard for die sinking EDM. The effect of parameters such as peak current, pulse on time, pulse off time and tool lift time were investigated during the EDM of precipitation hardening stainless steel with copper tungsten as tool electrode. Neural Network Toolbox in MATLAB was used to develop and assess the neural network model. The neural network model had two response variables which included MRR and SR. It was found that the difference between experimental data and predicted data was very low for both MRR and SR with average error as 3.32 and 2.25%. Figure 12 displays comparison of results for testing stage for MRR and SR respectively. The study corroborates the reliance on neural network model for estimating effect of parameters on die-sinker EDM process.

5 Optimization Techniques Used in Micro-EDM

Micro-EDM, developed from EDM, is used to machine parts in micro level. Micro-EDM has the same mechanism as the EDM process, with the differences in size of the tool electrode, axes movement and resolution, and level of discharge energy applied during machining. Similar to EDM process, micro-EDM also contains a series of variations, such as, micro-EDM drilling, micro die-sinking EDM, micro-EDM milling, micro wire-EDM (micro-WEDM) and micro WEDG (wire electro-discharge grinding). There were several optimization methods reported in the past years for optimization of the process parameters for improved micro-EDM performance. Depending on the method type, they could be categorized into three main methods: Taguchi Method, Grey Relational Method and Response Surface Method.

5.1 *Taguchi Method*

Maity et al. [55] applied Taguchi method to identify optimal process parameters of micro-EDM machine while fabricating micro-holes in copper using 300 μm diameter tungsten electrode. L9 orthogonal array design of experiment was applied. It was found that different combinations of optimum parameters could be obtained with different goals. When the goal of the machining was to reduce the machining time, the optimized settings for capacitance, feed rate, rpm and voltage were 0.1 μF, 0.003 μm/s, 1500 RPM and 90 V, respectively. When the goal of the machining was to minimize the recast layer thickness, the optimized settings for capacitance, feed rate, rpm and voltage were 0.0001 μF, 0.0003 μm/s, 1000 RPM and 110 V, respectively. When the goal of the machining was to find the optimized performance for all conditions, the optimized settings were suggested as 0.0001 μF, 0.001 μm/s, 1500 RPM and 120 V, respectively. In all the cases, the capacitance was found out to be the most influential factor in micro-EDM.

Azad et al. [56] used the Taguchi method in the micro-EDM drilling of titanium alloy using tungsten carbide electrode to optimize multiple performance characteristics of titanium alloy. The process parameters varied were pulse on time, frequency, voltage and supply current. And the performance characteristics studied were MRR, TWR and overcut (OC). The Taguchi orthogonal array L18 was found out to be the most suitable experimental design for this study. A confirmation test was also conducted. The Taguchi method was found to be an efficient experimental design method. The optimized process settings for micro-EDM drilling of titanium alloy were finally decided as 80 V voltage, 150 kHz pulse frequency, 1 A current and 50 μs pulse on time.

Kadirvel et al. [57] applied the Taguchi method to find optimized settings for obtaining a higher MRR, a lower TWR and the minimum SR in the micro die-sinking EDM process. During the study, the electrode was 300 μm silver tungsten (AgW) and the workpiece was EN-24 die steel. The Taguchi L16 orthogonal array method was applied under different conditions of gap voltage, capacitance, feed rate, and threshold voltage. After all the experiment, the ANOVA method was applied. The capacitance and the gap voltage were found to be the most significant independent variables influencing the performance characteristics during die-sinking micro-EDM process.

Chiou et al. [58] identified the optimized process parameters for the micro-EDM milling of high-speed steel alloy (SKH59) using tungsten carbide (WC) electrode. Taguchi method L9 orthogonal array was applied. The independent variables, i.e. process parameters, were decided as discharge current, pulse duration, pulse off time and jump distance. The performance characteristics were decided as TWR, MRR and overcut. Each independent variable contained three levels, from high to low. After the orthogonal array was set up, four columns and nine rows were contained in the array chart. The discharge energy was found to have the dominating effect on the three performance characteristics.

Lin et al. [7] applied the Taguchi method to optimize micro-EDM milling process parameters for machining of Inconel 718 alloy using 200 μm tungsten carbide electrodes. After all the Taguchi experiments were conducted, the signal-to-noise ratio (S/N) method was applied to determine the optimized settings based on the results. The S/N ratio has three different functions, larger is better, nominal is the best, and smaller is better. The goal of the study was to find the minimum electrode wear, the maximum material removal rate and the minimum working gap. By applying the Taguchi method, the experimental results showed the electrode wear was decreased by 7%, the MRR increased 357.5%, and the working gap decreased 6.3% in the optimum setting. For micro-EDM milling of Inconel 718 alloy, the optimized settings were found to be 0.5 A peak current, 3 μs pulse on time, 3 μs pulse off time and 60 V gap voltage. The peak current and gap voltage were the two factors that had the most influence on the performance characteristics.

5.2 Grey Relational Method (GRA)

Grey relational method is a data analysis method, which allows to find the grey relational grade of the parameters used in micro-EDM and identifies their effect on machining performance. Once all the experimental data are collected and the grey relational grades are calculated, the relationship between the process parameters and the performance parameters will be obvious. Bhosle et al. [13] used grey relational method to explore the optimized micro-EDM drilling conditions for machining micro-holes in Inconel 600 alloy using 500 μm diameter tungsten carbide electrodes. The performance characteristic considered were MRR, overcut, taper angle and diametric variance at entry and exit of micro-holes. Table 3 shows the Taguchi design of experiments used in this study and Table 4 presents the grey relation methods resulted from the analysis. The optimized micro-EDM drilling settings were suggested as 175 V voltage, 1000 pF capacitance, 20 μm/s EDM feed rate, 15 μs pulse duration and 50 μs pulse interval. It was found that the capacitance had the high-

Table 3 The design of experiments [13]

Exp. No.	A	B	C	D	E	MRR ($10^{-5}mm^3/s$)	Taper angle (°)	Overcut (μm)	DVEE
1	1	1	1	1	1	2.0401	6.2365	46.28	54.64
2	1	2	2	2	2	2.59	0.4079	50.24	3.56
3	1	3	3	3	3	14.34	0.8112	68.08	7.08
4	2	1	1	2	2	4.3412	3.3479	44.08	29.28
5	2	2	2	3	3	8.597	0.7562	54.88	6.6
6	2	3	3	1	1	33.912	0.7654	79.2	6.7
7	3	1	2	1	3	5.3179	1.6084	42.8	14
8	3	2	3	2	1	12.709	1.416	52.3	12.3
9	3	3	1	3	2	52.462	7.3798	87.2	64.8
10	1	1	3	3	2	1.1112	8.5442	35.44	75.12
11	1	2	1	1	3	4.1387	1.077	49.52	9.4
12	1	3	2	2	1	15.9932	5.2291	77.76	45.76
13	2	1	2	3	1	2.8099	2.5925	40.56	22.64
14	2	2	3	1	2	9.0418	0.3964	51.4	3.72
15	2	3	1	2	3	23.6342	1.2511	79.96	10.92
16	3	1	3	2	3	3.1063	1.0128	39.12	8.84
17	3	2	1	3	1	19.765	0.4629	53.2	4.04
18	3	3	2	1	2	46.2656	0.4766	78.48	1

est influence on performance characteristics followed by the voltage. The feed rate turned out to be the least influential. However, feed rate played important role in controlling the taper angle.

5.3 Response Surface Method

Different from GRA method, the response surface method (RSM) helps the researchers to build a mathematical model between the process parameters and performance parameters in micro-EDM process. Once the mathematical model is set up, the relationship between the parameters can be obtained. Tiwary et al. [59] used the RSM to explore the influence of the micro-EDM process parameters while machining of Ti-6Al-4V. During the study, the electrode was 300 μm brass and the workpiece was Ti-6Al-4V alloy. The process parameters were decided as pulse on time, peak current, gap voltage, and flushing pressure, and the performance characteristics were MRR, TWR, overcut (OC), and taper. With the help of the RSM, a mathematical model was built for these process parameters and performance parameters. In order to test all the possible combinations, a total of 31 experiments were conducted based

Table 4 The example Grey relation method results [13]

Exp. No.	Normalized experimental results				Grey relational coefficient				Grey relational grade
	MRR	Taper	Overcut	DVEE	MRR	Taper	Overcut	DVEE	
1	0.018	0.2832	0.7905	0.2861	0.3373	0.4109	0.7047	0.4118	0.4661
2	0.0287	0.9985	0.714	1	0.3398	0.997	0.6361	1	0.7432
3	0.2576	0.949	0.3693	0.9508	0.4024	0.9074	0.4422	0.9104	0.6656
4	0.0629	0.6377	0.833	0.6405	0.3479	0.5798	0.7496	0.5817	0.5647
5	0.1457	0.9558	0.6244	9575	0.3691	0.9)87	0.571	0.92)6	0.6951
6	0.6387	0.9547	0.1545	0.9561	0.5805	0.9)69	0.3716	0.9192	0.6970
7	0.0819	0.8512	0.8578	0.8541	0.3525	0.7706	0.7785	0.7741	0.6689
8	0.2258	0.8748	0.6742	0.8778	0.3924	0.7997	0.6054	0.8036	0.6502
9	1	0.1429	0	0.1442	1	3684	3333	0.3687	0.5176
10	0	0	1	0	0.3333	0.3333	1	0.3333	0.4999
11	0.0589	0.9164	0.7279	0.9183	3469	0.8567	0.6475	0.8595	0.6776
12	0.2898	0.4068	0.1823	0.4102	0.4131	0.4573	0.3794	0.4587	0.4271
13	0.033	0.7304	0.901	0.7333	0.3408	0.6496	0.8347	0.6521	0.6193
14	0.1544	1	0.6916	0.9977	0.3715	1	0.6185	0.9954	0.7463
15	0.4386	0.8951	0.1398	0.8971	0.471	0.8265	0.3675	0.8293	0.6235
16	0.0388	0.9243	0.9289	0.9262	3421	0.8685	0.8755	0.8713	0.7393
17	0.3632	0.9918	0.6568	0.9932	0.4398	0.9838	0.5929	0.9865	0.7507
18	0.8793	0.9901	0.1684	0.99)6	0.8055	0.9805	0.3754	0.9834	0.7862

on uniform rotatable central method. At the end of each experiment, all 4 of the performance characteristics were also recorded in the database. After all the experiments were conducted, the data were analyzed using "Minitab version 15" software. Based on the built mathematical model, all the coefficients were calculated by "Minitab version 15" for each performance characteristics. The average percentage of prediction errors for MRR, TWR, OC and taper were 2.61, 3.74, 3.21 and 3.7% respectively. The researchers also suggested that in order to achieve the optimized performance characteristics while machining Ti-6Al-4V alloy, i.e. the maximum MRR and the minimum TWR, OC and taper, the pulse on time should be set as 1 μs, the peak current should be set as 2.5 A, the gap voltage should be set as 50 V and the flushing pressure should be set as 0.2 kg/cm^2. Under those settings, the maximum MRR was 0.0777 mg/min, and the minimum values of TWR, OC, and taper were 0.0088 mg/min, 0.0765 mm, and 0.0013, respectively.

6 Conclusion and Future Research

Electrical discharge machining (EDM) usage has been drastically increasing in industrial applications. Therefore, optimization of machining parameters is essential to produce cost effective finished products. This chapter provides a concise overview of the various optimization techniques used in EDM, and entails comprehensive literature work conducted on optimization of EDM based processes using numerous statistical techniques. Die-sinking, fast hole drilling, wire, and micro EDM were used as machining processes with varieties of dielectric fluids and electrodes. Critical information on the findings of optimization in EDM has been covered in this chapter. Below are some of the key results from the chapter:

Wire EDM optimization:

- There have been more studies on optimizing the process parameters for wire EDM compared to die-sinking EDM and micro-EDM, which demonstrates more stochastic nature of the wire EDM process and importance of optimization process parameters in wire EDM.
- Taguchi method was found to be effective in designing the experiments. However, Taguchi based design of experiments are supported by other optimization techniques. ANOVA was used more commonly to identify the significance of each operating parameters on the machining performance.
- In many cases, combination of multiple optimization methods and comparing the outcome of different methods with each other may be an effective approach to identify the optimal process parameters and machining conditions in EDM.
- Among various methods, neural network and simulated annealing were found to be very accurate in predicting the optimum conditions, when compared with experimental results.

- The response surface method was found to be useful in developing empirical models and identifying the influence of individual parameter or combination of parameters on the performance characteristics of EDM.
- The peak current and pulse duration were found to be the two most significant parameters in wire EDM.

Sinker EDM optimization:

- Response surface method and Taguchi method were both found to be effective and recommended in determining optimal parameters and improving quality for sinker EDM optimization.
- Pulse on time, pulse current, and duty cycle were the most significant input parameters affecting material removal rate, surface roughness, tool wear rate, and over-cut during sinker EDM.
- Tool wear rate, material removal rate and over-cut generally increased with current and pulse on time. The electrical conductivity and thermal conductivity of workpiece and electrode material were also found to be significant in affecting output variables in sinker EDM.

Micro EDM optimization:

- The capacitance and the gap voltage were found to be the most significant independent variables influencing the performance characteristics for die-sinking micro-EDM.
- The peak current and spark gap were the two factors that had the most influence on the performance characteristics for micro-EDM milling.
- In micro-EDM drilling, the capacitance had the most influence on both side gap and taper ratio, followed by gap voltage. The feed rate turned out to be the least influential. However, feed rate plays important role in controlling the taper angle.
- The optimized settings depend on the goal of the experiment. Different optimized performance characteristics require different optimized process setting combinations.

New materials with intricate shapes and versatile properties are created frequently to maximize performance and production. Thus, constant effort should be made in the research field to comprehend the impact of machining parameters on the materials and productivity, and to find optimal parameters which can be used in industries. Following are some of the ideas and challenges that can be tackled in the future research:

- Limited research was conducted using various shapes or dimensions of electrodes in die-sinking EDM. Future work should comprise of electrodes with common shapes used in industries. Analysis of dielectric fluid in optimization was insufficient, and needs further exploration.
- Due to growing demand of composites, the optimization techniques in EDM of popular metal matrix composites (MMCs) should be studied.

- Responses such as residual stress, surface morphology, and mechanical properties after EDM must be analyzed moving forward.
- Cost analysis and mass production efficiency should also be studied in addition to optimization to provide pragmatic outcome of changes in parameters.
- After the initial experiments are finished, a repeated set of experiments or machining a final product with optimized settings could be conducted in order to guarantee the optimized results could be used in the future operations in industries.
- Depending on fields of study, the process parameters and the performance parameters could vary and new research should consider optimization in real industrial applications.

References

1. Gupta K, Gupta MK (2019) Developments in non-conventional machining for sustainable production-a state of art review. Proc IMechE Part C J Mech Eng Sci. https://doi.org/10.1177/0954406218811982
2. Jahan MP (2015) Electrical discharge machining (EDM) types, technologies and applications. Nova Science Publishers, Inc., New York
3. Jahan MP, Asad ABMA, Rahman M, Wong YS, Masaki T (2011) Micro-electro discharge machining (MEDM). In: Koç M, Ozel T (eds) Micro-manufacturing: design and manufacturing of micro-products. Willey, pp 301–346
4. Harris LN (1989) Taguchi techniques for quality engineering, Philip J. Ross, McGraw-Hill Book Company, 1988. Qual Reliab Eng Int 5(3):249–249
5. Huang JT, Liao YS (2003) Optimization of machining parameters of wire-EDM based on grey relational and statistical analyses. Int J Prod Res 41(8):1707–1720
6. Sankar, Chapter 2: Introduction to the Taguchi method. University of Massachusetts
7. Lin MY, Tsao CC, Hsu CY, Chiou AH, Huang PC, Lin YC (2013) Optimization of micro milling electrical discharge machining of Inconel 718 by Grey-Taguchi method. Trans Nonferrous Met Soc China (English Ed) 23(3):661–666
8. Ugrasen G, Ravindra HV, Prakash GVN, Keshavamurthy R (2014) Process optimization and estimation of machining performances using artificial neural network in wire EDM. Proc Mater Sci 6(Icmpc):1752–1760
9. Singh PN, Raghukandan K, Pai BC (2004) Optimization by Grey relational analysis of EDM parameters on machining Al-10%SiCP composites. J Mater Process Technol 155–156(1–3):1658–1661
10. Ikram A, Mufti NA, Saleem MQ, Khan AR (2013) parametric optimization for surface roughness, kerf and MRR in wire electrical discharge machining (WEDM) using Taguchi design of experiment. J Mech Sci Technol 27(7):2133–2141
11. Durairaj M, Sudharsun D, Swamynathan N (2013) Analysis of process parameters in wire EDM with stainless steel using single objective Taguchi method and multi objective Grey relational grade. Proc Eng, 868–877
12. Manna A, Bhattacharyya B (2006) Taguchi and Gauss elimination method: a dual response approach for parametric optimization of CNC wire cut EDM of PRAlSiCMMC. Int J Adv Manuf Technol 28(1–2):67–75
13. Bhosle RB, Sharma SB (2017) Multi-performance optimization of micro-EDM drilling process of Inconel 600 alloy. Mater Today Proc 4(2):1988–1997
14. Bobbili R, Madhu V, Gogia AK (2015) Multi response optimization of wire-EDM process parameters of ballistic grade aluminium alloy. Eng Sci Technol Int J 18(4):720–726

15. Kumar K, Ravikumar R (2014) Modeling and optimization of wire EDM process. Int J Mech Mechatron Eng IJMME-IJENS 14:454–457
16. Kanlayasiri K, Boonmung S (2007) Effects of wire-EDM machining variables on surface roughness of newly developed DC 53 die steel: design of experiments and regression model. J Mater Process Technol 192–193:459–464
17. Kuriakose S, Shunmugam MS (2005) Multi-objective optimization of wire-electro discharge machining process by non-dominated sorting genetic algorithm. J Mater Process Technol 170(1–2):133–141
18. Chen HC, Lin JC, Yang YK, Tsai CH (2010) Optimization of wire electrical discharge machining for pure tungsten using a neural network integrated simulated annealing approach. Expert Syst Appl 37(10):7147–7153
19. Yang S-HH, Srinivas J, Mohan S, Lee D-MM, Balaji S (2009) Optimization of electric discharge machining using simulated annealing. J Mater Process Technol 209(9):4471–4475
20. Jung JH, Kwon WT (2010) Optimization of EDM process for multiple performance characteristics using Taguchi method and Grey relational analysis. j Mech Sci Technol 24(5):1083–1090
21. Goswami A, Kumar J (2014) Optimization in wire-cut EDM of Nimonic-80A using Taguchi's approach and utility concept. Eng Sci Technol Int J 17(4):236–246
22. Ramakrishnan R, Karunamoorthy L (2006) Multi response optimization of wire EDM operations using robust design of experiments. Int J Adv Manuf Technol 29(1–2):105–112
23. Mahapatra SS, Patnaik A (2007) Optimization of wire electrical discharge machining (WEDM) process parameters using Taguchi method. Int J Adv Manuf Technol 34(9–10):911–925
24. Tilekar S, Das SS, Patowari PPK (2014) Process parameter optimization of wire EDM on aluminum and mild steel by using Taguchi method. Proc Mater Sci 5:2577–2584
25. Rajyalakshmi G, Venkata Ramaiah P (2013) Multiple process parameter optimization of wire electrical discharge machining on Inconel 825 using Taguchi Grey relational analysis. Int J Adv Manuf Technol 69(5–8):1249–1262
26. Raj DA, Senthilvelan T (2015) Empirical modelling and optimization of process parameters of machining titanium alloy by wire-EDM using RSM. Mater Today Proc 2(4–5):1682–1690
27. Sarkar S, Sekh M, Mitra S, Bhattacharyya B (2008) Modeling and optimization of wire electrical discharge machining of γ-TiAl in trim cutting operation. J Mater Process Technol 205(1–3):376–387
28. Sarkar S, Mitra S, Bhattacharyya B (2005) Parametric analysis and optimization of wire electrical discharge machining of γ-titanium aluminide alloy. J Mater Process Technol 159(3):286–294
29. Mahapatra SS, Patnaik A (2006) Parametric optimization of wire electrical discharge machining (WEDM) process using Taguchi method. J Braz Soc Mech Sci Eng 28(4):422–429
30. Kumar K, Agarwal S (2012) Multi-objective parametric optimization on machining with wire electric discharge machining. Int J Adv Manuf Technol 62(5–8):617–633
31. Rao PS, Ramji K, Satyanarayana B (2014) Experimental investigation and optimization of wire EDM parameters for surface roughness, MRR and white layer in machining of aluminium alloy. Proc Mater Sci 5:2197–2206
32. Spedding T, Wang Z (1997) Study on modeling of wire EDM process. J Mater Process Technol 69(1–3):18–28
33. Yang RT, Tzeng CJ, Yang YK, Hsieh MH (2012) Optimization of wire electrical discharge machining process parameters for cutting tungsten. Int J Adv Manuf Technol 60(1–4):135–147
34. Çaydaş U, Hasçalik A, Ekici S (2009) An adaptive neuro-fuzzy inference system (ANFIS) model for wire-EDM. Expert Syst Appl 36(3 PART 2):6135–6139
35. Tarng YS, Ma S, Chung LK (1995) Determination of optimal cutting parameters in wire electrical discharge machining. Int J Fatigue 35(12):1693–1701

36. Spedding TA, Wang ZQ (1997) Parametric optimization and surface characterization of wire electrical discharge machining process. Precis Eng 20(1):5–15
37. Balasubramanian P, Senthilvelan T (2014) Optimization of machining parameters in EDM process using cast and sintered copper electrodes. Proc Mater Sci 6(Icmpc):1292–1302
38. Mishra BP, Routara BC (2017) Comparative analysis of the performance effectiveness of sinking electrical discharge machining (EDM) process with enhancement in hardness of the parent EN-31 alloy steel. Mater Today Proc 4(9):10235–10239
39. Payaghan NS, Ubale SB, Kadam MS (2014) Analysis & optimization of machining parameters for Ra in EDM process of Cu-W metal matrix composite(MMC). Int J Mech Eng Inf Technol 2(6):477–485
40. Leao FN, Pashby IR, Cuttell M, Lord P (2005) Optimisation of EDM fast hole-drilling through evaluation of dielectric and electrode material. In: 18th international congress of mechanical engineering
41. Urade AD, Deshpande VS (2016) Experimental investigations of EDM process parameters for tool wear rate based on orthogonal array. Int J Sci Res Sci Eng Technol 5(5):415–419
42. Lee DH, Malhotra N, Jung DW (2017) Multi characteristic optimization in die sinking EDM of En31 tool steel using utility concept. In: 2017 8th international conference on mechanical and aerospace engineering, ICMAE 2017, pp 166–170
43. Malhotra N (2013) Optimization of multiple quality characteristics of EDM process for MRR and TWR using utility concept. Adv Prod Eng Manag 8(4):219–230
44. Rath U (2017) Parametric optimization of EDM on EN19 using Grey-Taguchi analysis. Indian J Sci Technol 10(23):1–7
45. Prasad L, Gupta A (2017) An experimental investigation of machining parameters for EDM using copper electrode of Aisi P20 tool steel. Asian J Sci Technol 08(01):4106–4111
46. Belgassim O, Abusada A (2012) Optimization of the EDM parameters on the surface roughness of AISI D3 tool steel. In: Proceedings of the 2012 international conference on industrial engineering and operations management, Istanbul, Turkey, July 3–6, pp 2306–2313
47. Shrivastava SM, Sarathe AK (2014) Experimental investigation of surface roughness of SS-202 in die sinking EDM. Retrieved from https://www.academia.edu/10482710/Experimental_Investigation_of_Surface_Roughness_of_SS-202_in_Die_Sinking_EDM (last accessed on Jan 19, 2019)
48. Panda MR, Sagar S, Panigrahi AK, Swain MK (2016) Experimental analysis and optimization of EDM on AA6061 using. Int J Eng Sci Res Technol 5(4):265–271
49. Shahril M, Yusop M, Azhar MA, Ramli AS, Hamimi I, Razak A (2017) Effect of spray and complete submerged method on material removal rate for EDM die sinking. Int J Appl Eng Res 12(24):973–4562
50. Sangeetha M, Reddy AS, Kumar GV (2017) Optimization of die-sinking EDM process parameters in machining of AMMC-desirability approach. Mech Mater Sci Eng J, 7
51. Dubey S, Agrawal P, Jha A, Tech Student M (2017) A comprehensive study of electro discharge machining of Al/Al_2O_3 composite and it's optimization. Int J Eng Sci Comput, 13963–13966
52. Gaikwad V, Jatti VS (2016) Optimization of material removal rate during electrical discharge machining of cryo-treated NiTi alloys using Taguchi's method. J King Saud Univ Eng Sci, 0–6
53. Daud MR, Samion S, Yahya A, Mahmud N, Liyana N, Hashim S, Nugroho K (2013) The effect of current rates on surface SKD 11 for electrical discharge machine die sinker. In: International conference on mechanical engineering research, Bukit Gambang Resort City, pp 1–3
54. Chandramouli S, Eswaraiah K (2017) Optimization of EDM process parameters in machining of 17-4 PH steel using Taguchi method. Mater Today Proc 4(2):2040–2047
55. Maity KP, Singh RK (2012) An optimisation of micro-EDM operation for fabrication of micro-hole. Int J Adv Manuf Technol 61(9–12):1221–1229
56. Azad MS, Puri AB (2012) Simultaneous optimisation of multiple performance characteristics in micro-EDM drilling of titanium alloy. Int J Adv Manuf Technol 61(9–12):1231–1239
57. Kadirvel A, Hariharan P (2014) Optimization of the die-sinking micro-EDM process for multiple performance characteristics using the Taguchi-based Grey relational analysis. Mater Tehnol 48(1):27–32

58. Chiou AH, Tsao CC, Hsu CY (2015) A study of the machining characteristics of micro EDM milling and its improvement by electrode coating. Int J Adv Manuf Technol 78(9–12):1857–1864
59. Tiwary AP, Pradhan BB, Bhattacharyya B (2014) Study on the influence of micro-EDM process parameters during machining of Ti–6Al–4V superalloy. Int J Adv Manuf Technol 76(1–4):151–160

Optimization of Accuracy and Surface Finish of Drilled Holes in 350 Mild Steel

A. Pramanik, A. K. Basak, M. N. Islam, Y. Dong, Sujan Debnath and Jay J. Vora

Abstract This chapter presents analysis and optimization of machinability of Mild steel grade 350 while high speed drilling operation. Taguchi design of experiments (DoEs), analysis of variance (ANOVA) and other traditional methods were applied to optimize the input variables in order to minimise the circularity, cylindricity, diameter error and surface roughness of drilled holes. It was found that point angle was the highest contributor for the circularity, cylindricity and surface roughness of drilled holes. The circularity error was minimum at the low speed (584 rpm), low feed (0.15 mm/rev) and moderate point angle (125°). The cylindricity error of holes was minimised at the high speed (849 rpm), moderate feed (0.2 mm/rev) and moderate point angle (125°). The moderate speed, low feed and moderate point angle minimised surface roughness considerably. The interaction between speed and point angle had the maximum contribution to the diameter error of drilled holes. The diameter error was minimum at the moderate speed, low feed and moderate point angle.

1 Introduction

Drilling is a widely used fabrication process to form cylindrical holes of varying size (diameters and depths) by removing materials with the help of rotating cutting tool called 'drill'. Drilling plays a very vital role in manufacturing industry, without

A. Pramanik (✉) · M. N. Islam · Y. Dong
School of Civil and Mechanical Engineering, Curtin University, GPO Box U1987, Perth, WA 6845, Australia
e-mail: alokesh.pramanik@curtin.edu.au

A. K. Basak
Adelaide Microscopy, The University of Adelaide, Adelaide, SA, Australia

S. Debnath
Department of Mechanical Engineering, Curtin University, Sarawak, Malaysia

J. J. Vora
Mechanical Engineering Department, School of Technology (SOT), Pandit Deendayal Petroleum University (PDPU), Gandhinagar 382007, India

K. Gupta and M. K. Gupta (eds.), *Optimization of Manufacturing Processes*, Springer Series in Advanced Manufacturing, https://doi.org/10.1007/978-3-030-19638-7_3

which no manufacturing/fabrication becomes possible. There are three regions in a drill tool, namely shank, flute and drill point. The function of a shank is to permit the drill to be held and driven. Flute can help to remove chips when the drill starts cutting the material, thus leading to the formation of chips of removed materials and heat generation due to the friction. Flute at that point assists in the removal of chips as well as carries cutting fluids all the way to the drill point. This is beneficial to extract the excessive heat form cutting zones and keep the operation running smoothly. The helix angle present in flute plays a very important role in the chip formation during a drilling process. Depending on the end-users' requirements, two different sets of helix angles can be chosen, namely low helix angle (15°–20°) and high helix angle (35°–40°). Low helix angle gives a greater torque and thrust due to its rigidity compared to low helix angle. High helix angle facilitates greater chip clearance for drilling a deeper hole in materials with less tensile strength. The point region of drill helps to penetrate workpiece materials. It also contains chisel edge and two cutting lips. For a standard drill bit, its point angle is generally 118° and a lip relief or clearance angle vary between 7 and 20° [12]. The quality of the hole fabricated by drilling operation in terms of cylindricity, circularity, diameter and surface finish depends on the values of input parameters. Inappropriate values can lead to an unnecessary increase of thrust and torques on cutting tool. In addition, vibrations produced by a spindle during cutting operation can result in increased thrust and torque. Also, it contributes to heat generation due to friction and cause workpiece deformation.

In this work, the type of workpiece taken into considerations for drilling is mild steel 350 as one of the most common metals used in industries for structural applications. Mild steel is most widely employed in applications with excellent mechanical strength and combinations of varying properties [3]. Due to its high hardness and malleability, it is used for construction materials, pipelines and cookware. It is easy to machine and shaped to required geometries because of its flexible behavior. To obtain the extra strength and hardness, it can always be hardened with the help of carburising and heat treatment. The hardness of this type of steel can reach up to 550 BHN. However, hardenability is restricted within thin areas close to surfaces due to the restricted diffusion of surface hardening elements during the process. As a result of high carbon content, mild steel is prone to rusting and therefore, preventive measures should be taken to avoid further rusting (https://sciencestruck.com/mild-steel-properties).

The main input parameters responsible for obtaining desired quality of drilling hole in a workpiece are cutting speed, cutting feed and point angle. Cutting speed is a machining distance per unit time covered by the cutting edge while the tool is rotating. Increase in speed results in higher frictional heat that induces tool wear and defects in the hole [8]. Cutting feed is defined as the rate at which tool is being penetrated in workpiece during drilling operation. It is one of the most important parameters in drilling operation. The failure to choose an appropriate value of this parameter may adversely affect the hole accuracy, burr formation, surface finish and drill breakage in worst case scenario [8]. Point angle of a drill is another cutting parameter of drilling operation. It is practically equivalent to the lead angle in turning and milling operations. The point angle of a drill results in surface penetration of a workpiece and giving rise to cutting forces. It is evident that more surfaces of a drill point come in contact to workpiece (decrease in point angle) with thinner and wider chip formation. Increases in cutting edge resulting in better tool life, reduction in axial forces and rise in radial thrust [12].

2 Literature Review

Optimization of any machining process is very important to improve the productivity and part/feature quality. There have been many investigations conducted to optimize the drilling process using various methods. For example, Lee et al. [24] used abductive network for modelling drilling processes where the network was composed of a number of functional nodes. These were self-organized and applied optimization by using a predicted squared error criterion. The developed network was capable to predict the drilling performance from the process parameters. After this, they applied simulated annealing optimization algorithm with a performance index to the developed network when searching for the optimal process parameters. Kim and Ramulu [19] utilized multiple objective linear program to optimize drilling feed and speed to maximize hole quality and minimize machining cost. The analysis of variance (ANOVA) technique was used to predict the relative significance of the process parameters and to estimate the experimental errors. ANOVA analysis of a model with input parameters and all the second order interactions were used to isolate important process factors. An empirical quadratic equation was used to determine the correlations between input and output parameters. Haq et al. [13] considered multiple responses based on orthogonal array with grey relational analysis to optimize drilling parameters. A grey relational grade is obtained from the grey analysis. Based on the grey relational grade, optimum levels of parameters have been identified and significant contribution of parameters is determined by ANOVA. Confirmation test is conducted to validate the test result. Experimental results have shown that the responses in drilling process can be improved effectively through the new approach. Kurt et al. [23] utilize Taguchi methods to optimize dry drilling process. Orthogonal arrays of Taguchi, the signal-to-noise (SNR) ratio, ANOVA and regression analyses are employed to find the optimal levels and to analyse the effect of the input parameters on output parameters. Krishnamoorthy et al. [21] used Taguchi's L_{27} orthogonal array where the optimal combination of drilling parameters was chosen using grey fuzzy relational analysis. The optimization of drilling parameters was based on five different output performance characteristics, namely, thrust force, torque, entry delamination, exit delamination and eccentricity of the holes. They used ANOVA to find the percentage contribution of the drilling parameters. Vankanti and Ganta [40] optimized drilling process based on Taguchi experimental design and an L9 orthogonal array to understand the effect of different combinations of input parameters on hole quality. ANOVA test was conducted to determine the significance of each process parameter on drilling. The Taguchi's experimental design and ANOVA techniques were also implemented by Prasanna et al.[32] to understand the effects, contribution, significance and optimal machine settings of process parameters during drilling small holes where grey relational analysis and mathematical modelling were undertaken by regression analysis.

The literature review shows that the drilling process has mainly been optimized by Taguchi's experimental design and ANOVA techniques. An abundant number of reports are available on drilling steels and many other materials in literature. Firouzdor et al. [11] investigated wear and tool life of HSS drills under dry machining of carbon steels. It was noted that tool life of the drill bits improved by 77 and 126% when drill bits underwent cryogenic and cryogenic-temper treatment, respectively. This resulted from fine and homogeneous carbide particles formation during the cryogenic treatment that deterred diffusion wear. In addition, improved hardness due to the martensite formation played an active role in the reduction of tool wear. Çiçek et al. [7] also used cryogenically treated M 35 high speed steel (HSS) twist drills in drilling AISI 304 and 316 stainless steels (SS). In this case, treated drills also performed better than untreated drills in terms of thrust generation, surface roughness, tool wear and tool life for both types of stainless steels. Tool life of treated HSS drills in drilling of AISI 304 SS and AISI 316 SS improved by 32 and 14%, respectively, compare to that of untreated drills. The machinability of AISI 304 SS was worse than that of AISI 316 SS. Dolinšek et al. [9] investigated the chip formation procedure using quick-stop during the process of drilling austenitic stainless steels. Severe work hardening was noted in cutting zone, predominantly owing to the action of chisel edge by feed (50%) and cutting (30%) forces. This formed a hard but narrow chisel edge chip acting as a built-up edge (BUE). Belluco et al. [2] evaluated the performance of vegetable-based oils while drilling AISI 316L austenitic stainless steel using conventional HSS-Co tools. They found that vegetable-based oils performed better than conventional mineral oil where the tool experienced as high as 177% increment of tool-life and as low as 7% reduction in thrust. Kıvak et al. [20] studied the optimisation of drilling parameters using Taguchi method and analysis of variance (ANOVA) during the dry drilling of AISI 316 stainless steel by uncoated and coated M35 HSS twist drills. It was found that the type of cutting tool was the most significant factor that affected surface roughness, whereas feed rate influenced the thrust most. It was reported that nano-TiAlN coated drill with the cutting speed of 18 m/min and feed rate of 0.12 mm/rev gave the best surface finish. On the other hand, uncoated drill with the cutting speed of 16 m/min and feed rate of 0.1 mm/rev generated the lowest thrust force. Chen et al. [6] investigated the performance of titanium nitride (TiN) and titanium carbonitride (TiCN) coated tools during the drilling of JIS SUS 304 SS. In this case, TiN multilayer coated drills performed better under different machining conditions than that of uncoated drill. Both thrust and torque increased with the rise of feed for both coated drills and increased with the rise of spindle speed up to 120 rpm, but decreased with further rise of speed to 157 rpm. The contributions of feed on thrust and torque for both coated drills were higher than that of spindle speed. Min et al. [25] developed a burr formation control chart during the drilling of low alloy steel AISI 4118, using high-speed steel split-point twist drills. Burrs were classified as small uniform burr, large uniform burr, crown burr, and transient burr based on their shapes. New parameters consisting of process parameters and drill diameter were developed and used to show a unique distribution of burr types. The uniform burrs were independent of drill diameter. There were certain limits of feed and cutting speed, above which crown burrs were generated. Routio et al. [34] investigated tool-monitoring methods during the drilling of stainless steel where three cutting fluids and high-speed steel (uncoated and TiN-coated) tools were compared. Nomani [27] investigated machinability of duplex alloys SAF 2205, SAF 2507 and austenite stainless steel 316L during the drilling process. It was observed that duplex alloys displayed poor machinability responses, with SAF 2507 being the worst where abrasion and adhesion wears appeared on flank and rake faces. The adhesion wear was most severe on the flank face that was triggered by built-up edge formation, which was the highest in case of duplex alloys. It was found that duplex 2507 had higher sensitivity to cutting speed during machining and strain hardening at a higher velocity and less machinability due to presence of higher percentage of Ni, Mo and Cr. the mechanism of BUE formation in stagnation region of duplex alloys [28, 29] was further investigated during the material removal by a turning process. Kilickap et al. [18] optimised input variables such as cutting speed, feed rate and point angle that affected burr height during the drilling of AISI 304 stainless steel by using the response surface methodology and genetic algorithm. It was reported that the minimum burr height could be achieved at lower cutting speed and feed rates, as well as at a higher point angle. Tosun [39] employed a grey relational analysis to optimise the drilling process parameters such as feed rate, cutting speed, drill point angles to minimize surface roughness and burr height. The order of importance with respect to controllable factors towards multi-performance characteristics was drill type, cutting speed, feed rate and drill point angle sequentially. It was claimed that surface roughness and burr height could be minimized via this approach. Hashmi et al. [14] used a fuzzy model to evaluate drilling parameters for different types of workpiece materials in a different range of hardness values. Çaydaş et al. [4] investigated the performances of HSS, K20 solid carbide and TiN-coated HSS tools in dry drilling AISI 304 austenitic stainless steel. Spindle speed, feed rate, drill point angle varied for different holes and surface roughness, tool flank wear, exit burr height and enlargement of hole size were measured. It was found that TiN-coated HSS drill performed the best with longer tool life and better hole quality, as well as lower surface roughness, which was followed by K20 carbide and HSS tools. Kaplan et al. [17] studied the effect of cutting speed and feed rate on cutting force and torque while dry drilling AISI D2 and AISI D3 cold work tool steels. It was determined that the most significant factor contributing to thrust was feed rate. High spindle speed increased the thermal expansion, resulting in the increased plastic deformation of the workpiece leading to the reduction in the microhardness of chips and further thrust reduction. Endo et al. [10] assessed the accuracy of drilled holes using a Fourier series approach for small holes of 1 mm diameter in a mild steel plate. The accuracy was measured by the extreme possible difference between hole's configuration and radius of least-square circle resulting in mean radius. The assessments by universal projector and a digital camera indicated that holes were not completely circular. They concluded that bending rigidity of drill tools significantly influenced the accuracy and led to an exceptional impact on drilled holes.

Hayajneh [15] investigated the effect of input parameters on the quality such as surface roughness, out-of-roundness and hole size of deep holes produced by boring trepanning association (BTA) drilling on medium carbon steel (AISI 1060). It was found that there was a critical cutting speed which gave the best surface finish. The roundness error increased with the increase of feed rate and cutting speed. The variation of hole was the minimum at the moderate feed rate, but it was higher at the low cutting speed. Samui [36] used MPMR (Minimax Probability machine regression), MARS (Multi-variate Adaptive Regression Spline) and LSSVM (Least square support vector machine) to investigate surface roughness and roundness error of drilled holes in AISI D2 workpiece by TiAlN/TiN multilayer coated cemented carbide drill bits. It was argued that all the methods remained valid for determining the output parameters and cutting tool significantly affected the roughness and roundness error. However, MPMR technique was found to be more accurate than MARS and LSSVM. Kumar et al. [22] used the Taguchi method to minimize the surface roughness of drilled holes in mild steel workpiece where the input variables were spindle speed, cutting feed and point angles. It was found that an integration of lower spindle speed (7 m/min), cutting feed (0.35 mm/rev) and intermediate point angle (90°) could be used to minimise the surface roughness. Sultan et al. [37] investigated the effect of cutting speed, feed rate and point angle on circularity, cylindricity, diameter error and surface roughness of drilled holes on AISI 316L stainless steel produced by uncoated carbide drill with 30° helix angle. It was reported that the effect of cutting speed on roundness error was minor, but the diameter error increased, and surface roughness decreased with the increase of cutting speed. The combined effect of cutting speed and feed rate on circularity error was not significant. The increase of cutting speed and feed rate increased the cylindricity error significantly, which further affected the straightness of the cutting tool.

3 Scope of the Investigation

The accuracy specification of machined holes is normally represented by their circularity, diameter error and cylindricity as the most important parameters. Circularity is also called roundness, which can be defined as the property of a surface on which all points are at an equal distance from the central axis of aforementioned surface [31]. It basically measures a deviation of geometric tolerance compared to a perfect circle. Figure 1 represents the schematic view of circularity of a particular hole.

Diameter error can be defined as the difference between experimental values obtained through coordinate measuring machine (CMM) and the actual size of the hole. If the diameter value obtained via CMM is greater than the actual size of the circle, it can be said that the hole has been over cut at a specified input parameter [31].

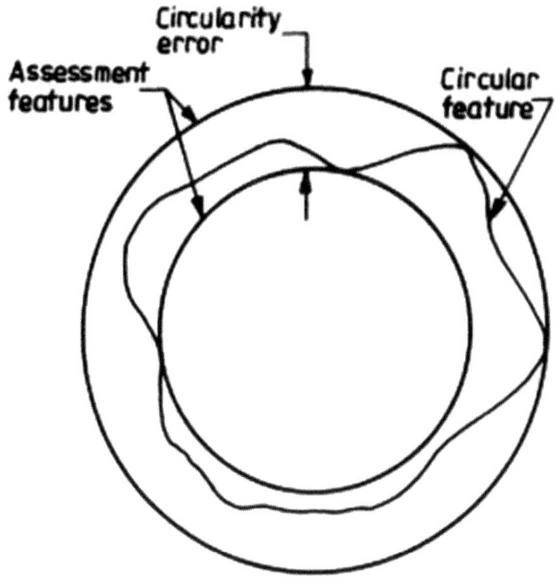

Fig. 1 Schematic of circularity error [35]

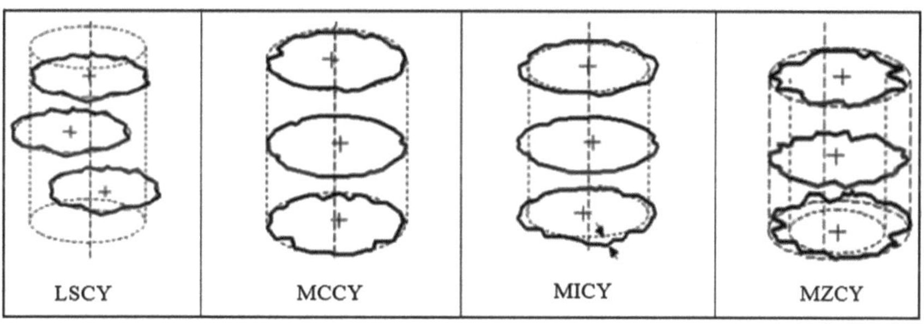

Fig. 2 Cylindricity analysis on a drilled hole [5]

Cylindricity is defined as a feature of a plane of revolution at which all points lying on the plane are at an equal distance from a common axis. Cylindricity tolerance can be explained as the determination of a tolerance zone for two concentric cylinders and planes where the cylinders must lie inside it [5]. In this case, the tolerance implies concurrently circular as well as longitudinal components of the surface. Cylindricity measurement incorporates circularity, straightness and tapering of a circular-hole dimension. In order to measure the cylindricity of a drilled hole, various methods can be employed, namely least square cylinder (LSCY), minimum circumcised cylinder (MCCY), maximum inscribed cylinder (MICY) and minimum zone cylinder (MZCY), as shown in Fig. 2.

Above-mentioned parameters accessing hole quality are very important in practical applications of drilled parts. When cylindrical fit is necessary, diameter error is the most important characteristic of a machined hole. It is particularly important for rotating component parts where excessive circularity may cause unacceptable vibration and heat. Another important quality parameter to examine is surface finish (e.g., surface roughness) as it is very significant in relation to wear, corrosion, fatigue, noise, load-carrying capacity, heat transfer and so on. Surface roughness represents random and repetitive deviations of a surface profile from nominal surface, which is usually expressed by arithmetic average (Ra) since it is most commonly used and internationally accepted.

The quality parameter data were analyzed by applying two statics techniques including Pareto ANOVA and Taguchi's signal-to-noise (SNR) ratio analysis through design of experiments (DOE). In a traditional ANOVA analysis, average response values have been used. This strategy is particularly suitable for monitoring trends or changes in the relationship of variables. However, it does not provide the complete representation because it normally does not include data with respect to the response scattering. Pareto ANOVA is a statistical method for determining the contribution of individual input parameters and their interactions with output quality parameters [30].

DOE is a powerful tool for experimentation which is widely used by researchers and engineers to study the effects of input parameters on output parameters in all fields. It plans experiments for appropriate data collection through the least number of experiments. Basically, DOE is the scientific management of information acquisition by experiment [30] which was first proposed by R. A. Fisher in England in the 1920s [26]. The inventive work dealt with agricultural applications of statistical methods. R. A. Fisher sought to find out how much rain, water, fertilizer, sunshine, etc. are needed to produce the best crop and pioneered the DOE methodology. It is also known as factorial DOE which can be full or partial. A full factorial DOE considers all possible combinations for a given set of factors and their levels. In general, full factorial DOE requires *nk* number of experimental runs, where *n* is the number of factor levels and *k* is the number of factors considered. The main advantage of this method is that it takes into account all the main and interaction effects, providing a full picture. However, the method requires a large number of experimental runs which are cumbersome, time consuming and expensive. The alternative is fractional factorial DOE in which only a small set of experiments are selected from a full factorial design. As a result, the interaction effects are often disregarded. While the fractional factorial method is well known, it is problematic as there are no guidelines for its application and subsequent analysis [16].

Taguchi formalized the fractional factorial DOE method and published a library of orthogonal arrays, which reduces the number of required experiments significantly. The method is simple and easy to apply. Orthogonality of the DOE permits the separation of the individual effects of each of several variables [38]. Taguchi's orthogonal array is represented in a symbolic format as $La(b^c)$, where the letter '*L*' indicates that the experimental designs are associated with Latin square designs, *a* is the number of runs, *b* is the number of levels considered, and *c* is the number

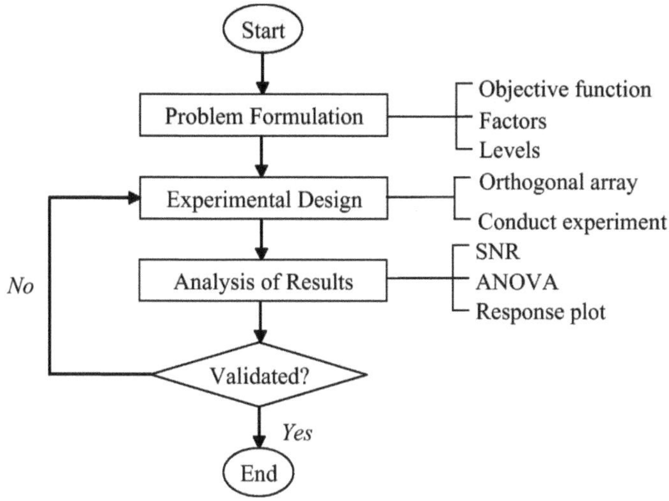

Fig. 3 Flow chart of Taguchi method [41]

of columns (number of factors) [16]. Thus, $L9$ (3^4) represents that a total of nine experimental runs will be conducted for a three-level, four-parameter experiment. A full factorial DOE conducted by the traditional method for the same study will need $3^4 = 81$ experiments. Hence, for an industrial application, the Taguchi method provides a significant savings of experimental runs.

The Taguchi method makes use of a special design of orthogonal array (OA) to examine the quality characteristics through a minimal number of experiments. The experimental results based on the OA are then transformed into S/N ratios to evaluate the performance characteristics [1]. The flowchart of the Taguchi method is illustrated in Fig. 3 [41]. It is more compact and includes a failure loop for invalidated design of experiment. The first step is the problem formulation, which requires defining an objective function, factors and levels. This step requires an in-depth knowledge of the process of interest. The inputs from experienced operators are essential in determining the potential optimization and disturbance variables, as well as their normal, high and low values. The optimization and disturbance variables are, respectively, called controllable and noise factors in the Taguchi-related literature. Limits of high and low values are called levels. Median, and lower and upper quartiles of the limits may also be included to augment experimental design configurations. The second step involves designing and conducting experiments. The number of factors and levels has an effect on selection of standard orthogonal arrays. The third step deals with the analysis of results. Three major statistical tools commonly applied in the Taguchi method are signal-to-noise ratio (SNR), analysis of means (ANOM) and analysis of variance (ANOVA). The fourth step is the validation of experiment.

In the traditional analysis, the average values of the response data are used, whereas the Taguchi method utilizes both average and variation of data. Therefore, the Taguchi method is expected to produce better results because it guarantees the highest quality with minimum variance [16]. Taguchi proposed the *S/N* ratio as a quantitative analysis tool for optimizing the outcome of a process. Taguchi classifies quality characteristics into three categories: (i) the smaller the better, (ii) the larger the better, and (iii) the nominal the better. The formula for calculating the *SNR* ratio depends on the type of quality characteristics investigated. For example, Eq. (1) calculates the *SNR* ratio of a quality characteristic in which the adage "the smaller the better" holds true [33].

To optimize the robustness of manufacturing process data, Taguchi statistical method was used by applying signal-to-noise ratio (SNR) was used in this investigation. The SNR can be calculated by using the following formula based on 'the smaller the better' criterion [33]:

$$SNR = -10\log\frac{1}{n}\left(\sum_{i=1}^{n} y_i^2\right) \quad (1)$$

where n is the number of observations and y is the observed data. The higher the value of S/N ratio, the better the results are because it guarantees the highest quality with a minimum variance. An expanded explanation of Taguchi method can be found elsewhere [33].

It is worth pointing out that the main emphasis of the Taguchi method is on robust design, i.e., making a product's quality of performance insensitive to variations in manufacture, in-service wear, and in-service environmental variations. It is a tool for quality improvement and cost reduction rather than determining the casual relationships of how things happen. The Taguchi approach is more engineering-oriented than science-oriented [30]. The results were analyzed applying average response, Taguchi's *S/N* ratio, and Pareto ANOVA. Pareto ANOVA is an excellent tool for determining the contribution of each input parameter and its interactions with the output parameters. It is a simplified ANOVA analysis method that does not require an ANOVA table and does not use F-tests. Consequently, it does not require detailed knowledge about the ANOVA method [30].

4 Experimental Procedure

Drilling process was carried out in a CNC machine centre as shown in Fig. 4. Experiments were performed on a metal plate (i.e., mild steel 350 with dimension of 440 × 75 × 12 mm and associated chemical compositions: 0.25% carbon, 1.5% manganese, 0.04% phosphorous, 0.04% sulphur and iron balance). Workpiece material including 12 mm thick and 12 mm diameter holes were drilled through the metal plate with the help of HSS drills with the specification of 12 mm diameter and different point angles (118°, 125° and 135°). Cutting fluid was used as coolant to minimize heat generation during machining. Circularity, diameter error and cylindricity of the holes

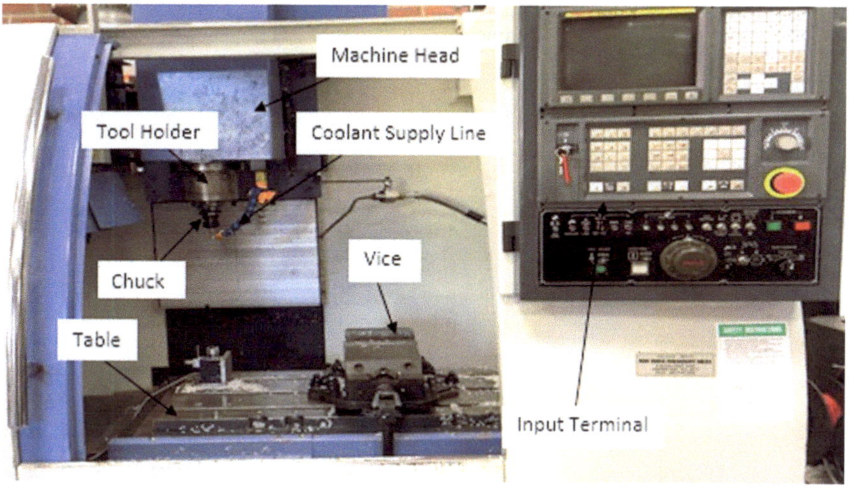

Fig. 4 Experimental setup

Table 1 Range and level of input parameters

Input parameters	Symbol	Level 0	Level 1	Level 2
Cutting speed (RPM)	A	584 (A0)	716 (A1)	849 (A2)
Feed (mm/rev)	B	0.15 (B0)	0.2 (B1)	0.25 (B2)
Point angle (°)	C	118° (C0)	125° (C1)	135° (C2)

were measured via coordinate measuring machine (CMM). A Discovery Model D-8 coordinate measuring machine (CMM), was used to determine the precision of cut hole, using Renishaw probe in a star configuration for convenience. The diameters of cut holes were determined by using standard built-in software of CMM. Eight coordinate points were measured to establish diameter values and each coordinate measurement was repeated three times. This coordinate data was also used to find the circularity of cut holes. Surface roughness of each turned surface was determined by using a tally surf Surftest SJ-201P. The measurement was taken at three locations on each hole with average data being reported.

These input parameters such as speed, feed and point angle were designed with the help of Taguchi method, as shown in Tables 1 and 2. The input parameters were categorized on the basis of different levels according to a L_{27} Taguchi approach. The levels were numbered starting from 0 to 2, where 0 was the lowest level while 2 represented the highest level.

Table 2 Details of each experiment

Experiment	Cutting speed (A)	Feed rate (B)	Point angle (C)
1	0	0	0
2	0	0	1
3	0	0	2
4	0	1	0
5	0	1	1
6	0	1	2
7	0	2	0
8	0	2	1
9	0	2	2
10	1	0	0
11	1	0	1
12	1	0	2
13	1	1	0
14	1	1	1
15	1	1	2
16	1	2	0
17	1	2	1
18	1	2	2
19	2	0	0
20	2	0	1
21	2	0	2
22	2	1	0
23	2	1	1
24	2	1	2
25	2	2	0
26	2	2	1
27	2	2	2

5 Results and Discussions

The optimization of input parameters such as speed, feed and point angle was based on the effect of these parameters on circularity, diameter error, cylindricity and surface roughness. The following results and discussions were presented according to the output parameters sequentially.

5.1 Circularity

The contributions of individual parameters and their interactions on the circularity of drilled holes were given in Table 3 based on Pareto ANOVA analysis. The table shows that the highest and most significant contributor on circularity was point angle (C), which contributed approximately 85% (P $\cong$ 85.78%). The next highest interacting contribution was from speed and point angle (A $\times$ C), approximately 3.5% (P $\cong$ 3.48). A very similar contribution (P $\cong$ 3.47) was also made by cutting speed (A). It was noted that individual parameters were most responsible factors affecting the circularity with 89.90%. Whereas, interactions of parameters only affected circularity by 10.10%. This made it easier to analyze that input parameters lead to the reduction in the circularity of drilled holes.

In order to validate the results obtained from Pareto ANOVA analysis, response table and graph for the circularity are presented in Table 4 and Fig. 5, respectively. These depicted input parameters and their responses against S/N ratio. The slopes obtained in the graphs express potential influence ratio that can further be valued. As seen from Table 4, it was noted that the lowest individual parameter contributing to higher circularity was point angle (C). This was followed by the contributions of speed and point angle interaction (A $\times$ C) and cutting speed (A). Hence, it can be said that in order to minimize the circularity, A_0B_0C1 was regarded as the perfect set of input parameters, where A_0 was lowest cutting speed (584 RPM), B_0 was low feed rate (0.15 mm/rev) and C_1 was moderate point angle (125°). The above-stated results can be further verified using a traditional method, as shown in Fig. 6. Thus it is confirmed that the minimum circularity can be found with the lowest cutting speed, lowest feed rate and moderate point angle. This can also be proven by using Taguchi analysis with similar results. Based on the consideration of input parameters (i.e., cutting speed, feed and point angle) obtained from ANOVA, Taguchi and traditional methods, it can be confirmed that all three methods yield similar results to optimized parameters leading to the minimum circularity.

5.2 Cylindricity

The contributions of input parameters and their interactions on the cylindricity of drilled holes are given in Table 5 based on Pareto ANOVA analysis. It was found that the most significant individual contributing parameter was point angle (C) where the influence ratio was P $\cong$ 40.8%. This was followed by the contribution of interaction effect between cutting speed and point angle (A $\times$ C) and then interaction effect between feed rate and point angle (B $\times$ C) with influence ratios of P $\cong$ 14.45 and 12.86%, respectively. In this case, contributions of interactions among the parameters (P $\cong$ 56.62%) on the cylindricity were greater than that of individual parameters (P $\cong$ 43.38%). This made it more difficult to analyze the input parameters leading to the

Table 3 Pareto ANOVA data for circularity

Sum at factor level	Factor and interaction								
	A	B	A × B	A × B	C	A × C	A × C	B × C	B × C
0	242.36	236.07	224.34	230.14	234.83	221.71	226.53	223.60	234.40
1	225.79	230.61	237.52	228.98	276.16	240.15	231.50	232.89	234.48
2	226.77	228.24	233.06	235.80	183.94	233.06	236.90	238.43	226.04
Sum of squares of difference (S)	518.44	96.73	269.51	79.88	12804.89	519.09	161.54	336.93	141.28
Contribution ratio (%)	3.47	0.65	1.81	0.54	85.78	3.48	1.08	2.26	0.95
Cumulative contribution	86.12	89.78	92.98	94.93	96.09	97.20	98.22	99.14	100.00
Check on significant interaction	A × C two-way table								
Optimum combination of significant factor level			A0B0C1						

Table 4 Response table for mean S/N ratio in relation to circularity and significant interaction

Input parameter	Symbol	Level 0	Level 1	Level 2	max-min
Speed	A	26.93	25.09	25.20	1.84
Feed	B	26.23	25.62	25.36	0.87
Point angle	C	26.09	30.68	20.44	10.25
Interaction	A × C	24.63	26.68	25.90	2.05

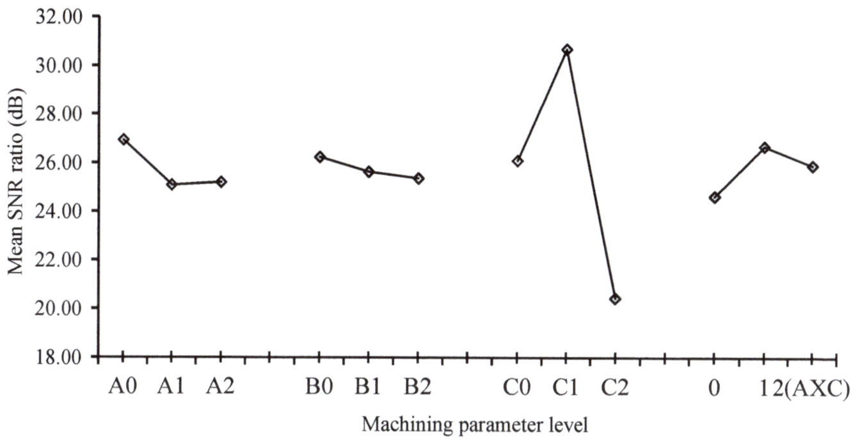

Fig. 5 Response graph of SNR ratio for circularity

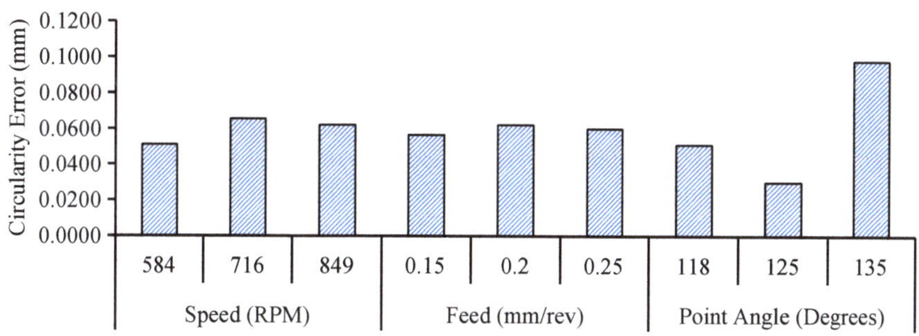

Fig. 6 Impact of input parameters on circularity using traditional method

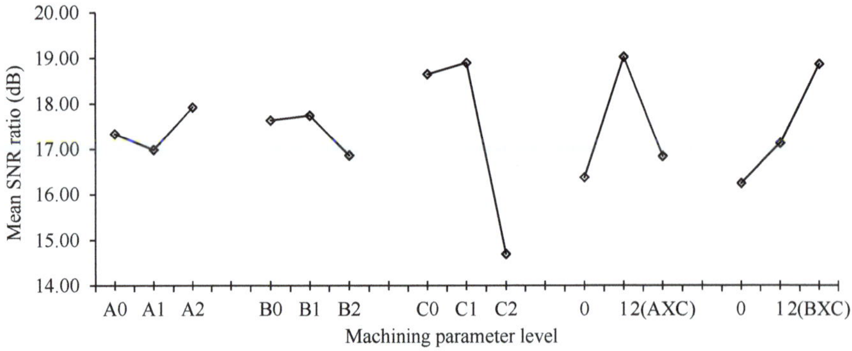

Fig. 7 Response graph SNR ratio for cylindricity

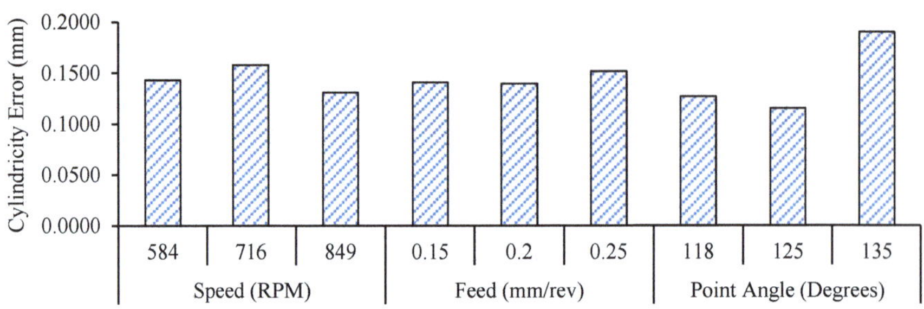

Fig. 8 Effect of input parameters on cylindricity using traditional method

least cylindricity error in holes, as interaction parameters involved a two-way table in order to evaluate the influence of individual parameters on the cylindricity.

Table 6 and Fig. 7 respectively present SNR response and graph for the cylindricity to validate the results obtained from Pareto ANOVA analysis. The slopes obtained in the graphs expressed the potential of influence ratio that could further be confirmed using Table 6. As shown from Table 6, it was noted that the most remarkable individual parameter contributing to highest cylindricity was point Angle (C). This was followed by feed rate (B) and cutting speed (A). As also illustrated in Fig. 7, it was found that the optimum parameters contributing to minimum cylindricity error were $A_2B_1C_1$, where A_2 was maximum cutting speed (849 RPM), B_2 was moderate feed rate (0.2 mm/rev) and C_2 was moderate point angle (125°). The above-stated results were further proven by using the traditional method, as displayed in Fig. 8. It was also showed that the minimum cylindricity error of drilled holes occurred at maximum cutting speed, moderate feed rate and moderate point angle. Therefore, the values of optimum input parameters (i.e., cutting speed, feed and point angle) obtained from ANOVA, Taguchi and traditional methods are similar in order to minimise the cylindricity error of drilled holes.

Table 5 Pareto ANOVA for cylindricity

Sum at factor level	Factor and interaction								
	A	B	A × B	A × B	C	A × C	A × C	B × C	B × C
0	156.02	158.74	144.71	151.47	167.84	147.39	152.07	151.49	146.21
1	152.86	159.69	163.67	153.68	170.06	171.22	149.39	165.32	154.18
2	161.32	151.76	161.81	165.04	132.30	151.58	168.74	153.39	169.81
Sum of squares of difference (S)	109.68	112.56	654.88	318.16	2693.79	970.96	659.38	337.40	864.57
Contribution ratio (%)	1.63	1.67	9.74	4.73	40.08	14.45	9.81	5.02	12.86
Cumulative contribution	40.08	54.53	67.39	77.20	86.94	91.96	96.70	98.37	100.00
Check on significant interaction			AXB two-way table						
Optimum combination of significant factor level			A2B1C1						

Table 6 Response table for mean SNR ratio for cylindricity

Input parameter	Symbol	Level 0	Level 1	Level 2	max-min
Speed	A	17.34	16.98	17.92	0.94
Feed	B	17.64	17.74	16.86	0.88
Point angle	C	18.65	18.90	14.70	4.20
Interaction	A × C	16.38	19.02	16.84	2.65

5.3 *Diameter Error*

The contributions of different input parameters and their interactions on diameter error of the holes are presented in Table 7 based on Pareto ANOVA analysis. It was shown that the highest contributor on diameter error was the interaction between cutting speed (A) and point angle (C) with an influence ratio of $P \cong 44.78\%$. This was followed by the contribution of second interaction between the same parameters (C and A) with the influence ratio of $P \cong 27.16\%$. The third highest contributor was point angle which had the influence ratio of $P \cong 16.73\%$. Based on total contributions from individual and interaction of parameters, it was noted that individual parameters contributed on diameter error by $P \cong 23.02\%$ whereas the contributions from the interaction of parameters was $P \cong 76.98\%$. Therefore, it was not easier to minimise diameter error by simply controlling individual parameters.

The results obtained from Pareto ANOVA analysis regarding diameter error were validate by SNR response table and graph, as demonstrated in Table 8 and Fig. 9, respectively. As the slopes of the graphs expressed the potential of influence ratio, Table 8 showed that the highest individual contributor was point angle (C) and lowest individual contributor was feed (B). Figure 9 showed that the minimum diameter error occurred at the lowest cutting speed ($A_0 = 584$ RPM), moderate feed rate ($B_1 = 0.20$ mm/rev) and lowest point angle ($C_0 = 118°$). These results were further validated by using the traditional method as depicted in Fig. 10. It was clearly demonstrated that the lowest diameter error took place with the lowest cutting speed, moderate feed rate and lowest point angle.

5.4 *Surface Roughness*

Based on Pareto ANOVA analysis, the contributions of different input parameters on surface roughness (Ra) of drilled holes were presented in Table 9. It was shown that the highest contributor to surface roughness of the drilled holes was point angle (C), which was followed by cutting speed (A). The contributions of these two parameters were $P \cong 48.6$ and 12.5%, respectively. These were followed by the contributions of interactions between feed rate and point angle (B × C) and the other interaction

Table 7 Pareto ANOVA data for diameter error

Sum at factor level	Factor and interaction								
	A	B	A × B	A × B	C	A × C	A × C	B × C	B × C
0	138.61	126.92	122.46	127.68	145.27	129.41	118.36	125.95	124.43
1	124.25	131.09	132.68	128.13	121.40	151.86	118.28	127.53	131.10
2	123.60	128.44	131.32	130.64	119.79	105.18	149.82	132.98	130.93
Sum of squares of difference (S)	432.22	26.71	184.99	15.30	1222.42	3270.87	1984.32	81.47	86.60
Contribution ratio (%)	5.92	0.37	2.53	0.21	16.73	44.78	27.16	1.12	1.19
Cumulative contribution	44.78	71.94	88.68	94.60	97.13	98.31	99.43	99.79	100.00
Check on significant interaction			A × C and A × B two-way table						
Optimum combination of significant factor level			A0B1C0						

Table 8 Response table for mean SNR ratio for diameter error

Input parameter	Symbol	Level 0	Level 1	Level 2	max-min
Speed	A	15.40	13.81	13.73	1.67
Feed	B	14.10	14.57	14.27	0.46
Point Angle	C	16.14	13.49	13.31	2.83
Interaction	A × C	14.38	16.87	11.69	5.19

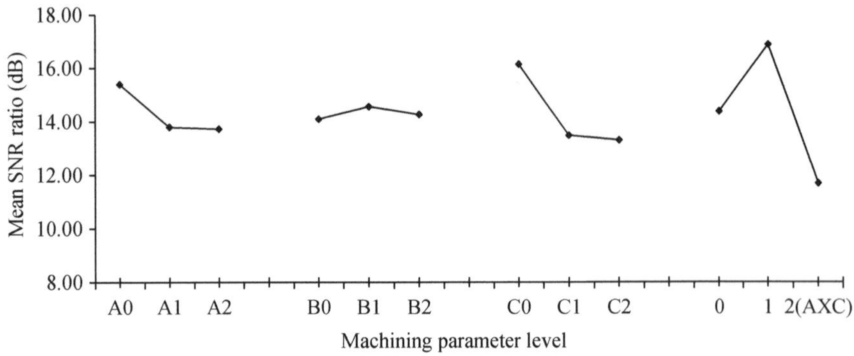

Fig. 9 Response graph SNR ratio for diameter error

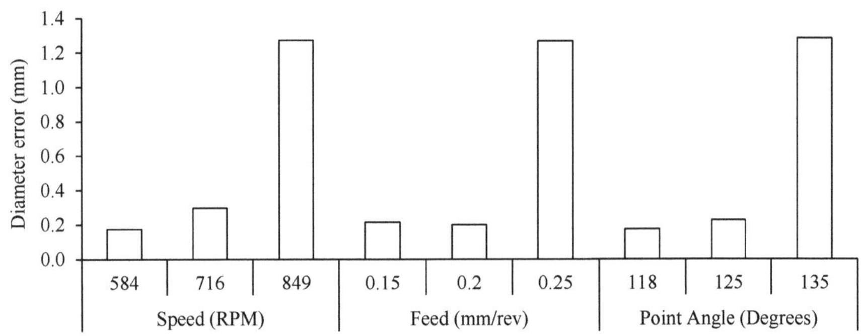

Fig. 10 Effect of input parameters on diameter error using traditional method

between cutting speed and point angle (A × C) with the influence ratios of P ≅ 10.85 and 9.97%, respectively. In this case, the total contribution from the effect of individual parameters was P ≅ 69.39%, which was much greater than that of interactions among parameters (P ≅ 30.61%). This made it easier to analyze the individual input parameters leading to the minimum surface roughness of drilled holes.

Table 10 and Fig. 11 represent the response of S/N ratio for surface roughness. Similar to Pareto ANOVA analysis, the response table showed that point angle (C) was the most contributing parameter for surface roughness. This was followed by

Table 9 Pareto ANOVA analysis for surface roughness

Sum at factor level	Factor and interaction								
	A	B	A x B	A x B	C	A x C	A x C	B x C	B x C
0	−154.45	−148.21	−163.29	−161.92	−170.42	−154.01	−168.72	−159.74	−149.84
1	−150.32	−163.47	−150.57	−157.62	−132.86	−156.39	−148.31	−157.54	−156.19
2	−171.90	−165.01	−162.82	−157.14	−173.39	−166.28	−159.65	−159.40	−170.65
Sum of squares of difference (S)	787.27	517.52	311.97	41.63	3062.43	253.78	627.43	8.43	682.62
Contribution ratio (%)	12.51	8.22	4.96	0.66	48.66	4.03	9.97	0.13	10.85
Cumulative contribution	48.66	61.17	72.02	81.99	90.21	95.17	99.20	99.87	100.00
Check on significant interaction			B × C and A × C two-way table						
Optimum combination of significant factor level			A1B0C1						

Table 10 Response table for mean S/N ratio for surface roughness

Input parameter	Symbol	Level 0	Level 1	Level 2	max-min
Speed	A	−17.16	−16.70	−19.10	2.40
Feed	B	−16.47	−18.16	−18.33	1.87
Point angle	C	−18.94	−14.76	−19.27	4.50
Interactions	A × C	−17.11	−17.38	−18.48	1.36

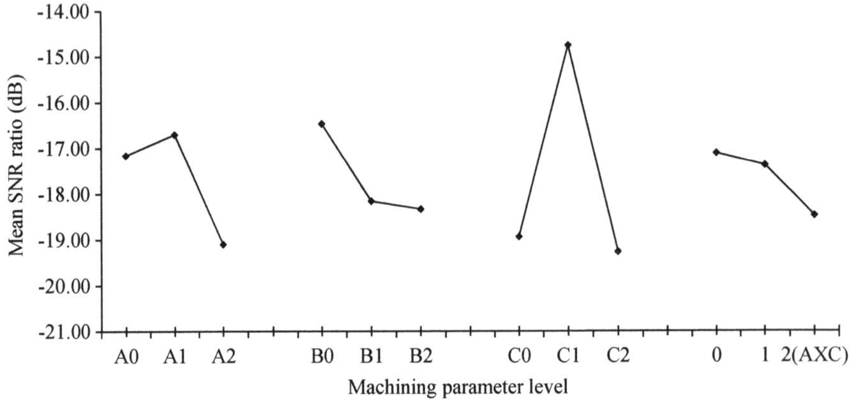

Fig. 11 Response graph of SNR ratio for surface roughness

cutting speed (A) and feed rate (B). As shown in Fig. 11, it was noted that optimum values of parameters contributing to the minimum surface roughness were $A_1B_0C_1$ where A_1 is moderate cutting speed (716 RPM), B_0 is lowest feed rate (0.15 mm/rev) and C_1 is moderate point angle (125°). The above-mentioned results were further proven using the traditional method, as shown in Fig. 12. It was also confirmed that the lowest surface roughness of drilled holes was found at moderate cutting speed, lowest feed rate and moderate point angle. As such, taking into consideration the values of input parameters (i.e., cutting speed, feed and point angle) obtained from ANOVA, Taguchi and traditional methods, it can be confirmed that all three methods gave similar results for optimised parameters leading to the least surface roughness of drilled holes.

6 Optimised Conditions

The optimised conditions of different output parameters are given in the Table 11. For the same conditions the CMM measurement of circularity error, cylindricity error and diameter error are presented in Fig. 13.

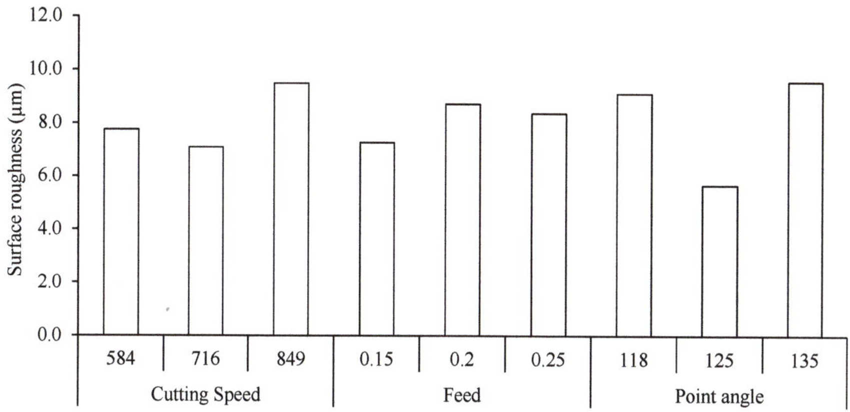

Fig. 12 Impact of input parameters on surface roughness using traditional method

Table 11 Values of input parameters for optimum output parameters

Output parameters	Speed (RPM)	Feed (mm/rev)	Point angle (°)
Circularity error A0B0C1	584	0.15	125
Cylindricity error A2B1C1	849	0.2	125
Diameter error A0B1C0	584	0.2	118
Surface roughness (A1B0C1)	716	0.15	125

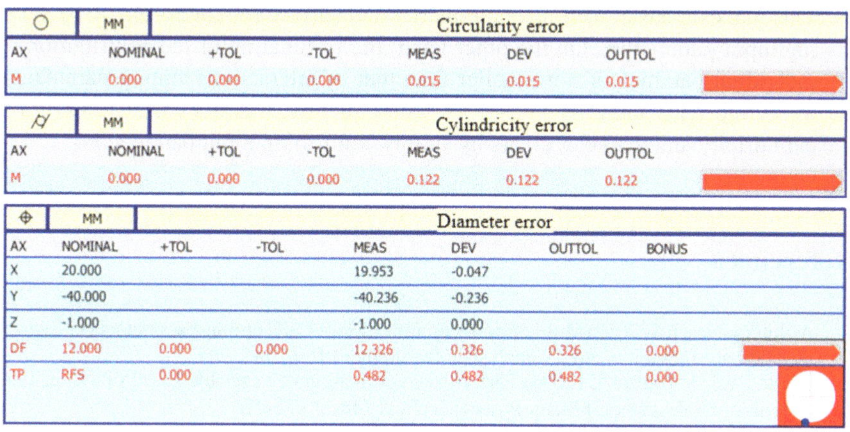

Fig. 13 CMM results for different error measurement at the optimized conditions

7 Conclusions

The accuracy and surface roughness of holes were analyzed by using Taguchi and Pareto ANOVA method while drilling holes in mild steel 350 grade. The accuracy of drilled holes has been evaluated based on three criteria: circularity, cylindricity and diameter error. The above-stated investigation on these factors can be summarized as follows:

1. Point angle was the most dominant factor to affect the circularity along the length of drilled holes, as confirmed by both Taguchi and ANOVA methods. Circularity error was the minimum at the low speed, low feed and moderate point angle. Minimum cylindricity error was achieved at the high speed, moderate feed and moderate point angle.
2. The interaction of speed and feed was the most significant factor to influence the diameter error of drilled holes. Low cutting speed, moderate feed and low point angle resulted in least diameter error. The diameter error for moderate cutting speed, low feed and moderate point angle was small as well, which was almost close to the parameters leading to lowest diameter error. Point angle was the most influencing factor, followed by cutting speed and interaction of feed and point angle as it caused the highest surface roughness drilled holes. Moderate speed, low feed and moderate point angle gave the minimum roughness along the length of drilled holes.
3. In case of circularity and surface roughness, the summation of the contribution of individual parameters is higher than that of interactions among parameters. This makes it easier to control circularity and surface roughness easily by varying input parameters. On the other hand, the summation of the contribution of individual parameters was smaller than that of interactions among parameters for cylindricity and diameter errors. Consequently, it is not easier to optimize cylindricity and diameter errors by simply controlling input parameters.

References

1. Asiltürk I, Neşeli S (2012) Multi response optimisation of CNC turning parameters via Taguchi method-based response surface analysis. Measurement 45(4):785–794
2. Belluco W, De Chiffre L (2004) Performance evaluation of vegetable-based oils in drilling austenitic stainless steel. J Mater Process Technol 148(2):171–176
3. Brady GS et al (2002) Materials handbook: an encyclopedia for managers, technical professionals, purchasing and production managers, technicians and supervisors. McGraw-Hill handbooks. McGraw-Hill
4. Çaydaş U et al (2011) Performance evaluation of different twist drills in dry drilling of AISI 304 austenitic stainless steel. Mater Manuf Processes 26(8):951–960
5. Chajda J et al (2008) Coordinate measurement of complicated parameters like roundness, cylindricity, gear teeth or free-form surface. In: International conference of advanced manufacturing operations, Kranevo, 8
6. Chen W-C, Liu X-D (2000) Study on the various coated twist drills for stainless steels drilling. J Mater Process Technol 99(1):226–230

7. Çiçek A et al (2012) Performance of cryogenically treated M35 HSS drills in drilling of austenitic stainless steels. Int J Adv Manuf Technol 60(1):65–73
8. Clyde FC (2001) Coombs' printed circuits handbook. McGraw-Hill Professional
9. Dolinšek S (2003) Work-hardening in the drilling of austenitic stainless steels. J Mater Process Technol 133(1):63–70
10. Endo H et al (2007) Accuracy estimation of drilled holes with small diameter and influence of drill parameter on the machining accuracy when drilling in mild steel sheet. Int J Mach Tools Manuf 47(1):175–181
11. Firouzdor V et al (2008) Effect of deep cryogenic treatment on wear resistance and tool life of M2 HSS drill. J Mater Process Technol 206(1):467–472
12. Geng H (2015) Manufacturing engineering handbook. McGraw Hill Professional
13. Haq AN et al (2008) Multi response optimization of machining parameters of drilling Al/SiC metal matrix composite using grey relational analysis in the Taguchi method. Int J Adv Manuf Technol 37(3–4):250–255
14. Hashmi K et al (2000) Fuzzy logic based data selection for the drilling process. J Mater Process Technol 108(1):55–61
15. Hayajneh MT (2001) Hole quality in deep hole drilling. Mater Manuf Processes 16(2):147–164
16. Islam MN, Pramanik A (2016) Comparison of design of experiments via traditional and Taguchi method. J Adv Manuf Syst 15(03):151–160
17. Kaplan Y et al (2014) Investigation of the effects of machining parameters on the thrust force and cutting torque in the drilling of AISI D2 and AISI D3 cold work tool steels
18. Kilickap E, Huseyinoglu M (2010) Selection of optimum drilling parameters on burr height using response surface methodology and genetic algorithm in drilling of AISI 304 stainless steel. Mater Manuf Processes 25(10):1068–1076
19. Kim D, Ramulu M (2004) Drilling process optimization for graphite/bismaleimide–titanium alloy stacks. Compos Struct 63(1):101–114
20. Kıvak T et al (2012) Taguchi method based optimisation of drilling parameters in drilling of AISI 316 steel with PVD monolayer and multilayer coated HSS drills. Measurement 45(6):1547–1557
21. Krishnamoorthy A et al (2012) Application of grey fuzzy logic for the optimization of drilling parameters for CFRP composites with multiple performance characteristics. Measurement 45(5):1286–1296
22. Kumar D et al (2012) Operational modeling for optimizing surface roughness in mild steel drilling using Taguchi technique. Int J Res Manag 2(3):66–77
23. Kurt M et al (2009) Application of Taguchi methods in the optimization of cutting parameters for surface finish and hole diameter accuracy in dry drilling processes. Int J Adv Manuf Technol 40(5–6):458–469
24. Lee B et al (1998) Modeling and optimization of drilling process. J Mater Process Technol 74(1–3):149–157
25. Min S et al (2001) Development of a drilling burr control chart for low alloy steel, AISI 4118. J Mater Process Technol 113(1):4–9
26. Montgomery DC (2017) Design and analysis of experiments. Wiley
27. Nomani J et al (2013) Machinability study of first generation duplex (2205), second generation duplex (2507) and austenite stainless steel during drilling process. Wear 304(1):20–28
28. Nomani J et al (2016) Investigation on the behavior of austenite and ferrite phases at stagnation region in the turning of duplex stainless steel alloys. Metall Mater Trans A 47(6):3165–3177
29. Nomani J et al (2017) Stagnation zone during the turning of Duplex SAF 2205 stainless steels alloy. Mater Manuf Process (just-accepted)
30. Park S (1996) Robust design and analysis for quality engineering. Boom Koninklijke Uitgevers
31. Pramanik A et al (2016) Accuracy and finish during wire electric discharge machining of metal matrix composites for different reinforcement size and machining conditions. Proc Inst Mech Eng Part B J Eng Manuf, 0954405416662079
32. Prasanna J et al (2014) Optimization of process parameters of small hole dry drilling in Ti–6Al–4V using Taguchi and grey relational analysis. Measurement 48:346–354

33. Ross PJ (1996) Taguchi techniques for quality engineering. McGraw-Hill International Editions
34. Routio M, Säynätjoki M (1995) Tool wear and failure in the drilling of stainless steel. J Mater Process Technol 52(1):35–43
35. Samuel G, Shunmugam M (2003) Evaluation of circularity and sphericity from coordinate measurement data. J Mater Process Technol 139(1–3):90–95
36. Samui P (2014) Determination of surface and hole quality in drilling of AISI D2 cold work tool steel using MPMR, MARS and LSSVM. J Adv Manuf Syst 13(04):237–246
37. Sultan A et al (2015) Chip formation when drilling AISI 316L stainless steel using carbide twist drill. Proc Manuf 2:224–229
38. Taguchi G (1987) System of experimental design: engineering methods to optimize quality and minimize cost. UNIPUB/Kraus Int. Pub, White Plains, NY
39. Tosun N (2006) Determination of optimum parameters for multi-performance characteristics in drilling by using grey relational analysis. Int J Adv Manuf Technol 28(5–6):450–455
40. Vankanti VK, Ganta V (2014) Optimization of process parameters in drilling of GFRP composite using Taguchi method. J Mater Res Technol 3(1):35–41
41. Yusoff N et al (2011) Taguchi's parametric design approach for the selection of optimization variables in a refrigerated gas plant. Chem Eng Res Des 89(6):665–675

Modelling and Optimization of Laser Additive Manufacturing Process of Ti Alloy Composite

Rasheedat M. Mahamood and Esther T. Akinlabi

Abstract Laser metal deposition process is one of the important processes of additive manufacturing technology which is used for the production of end-use parts as well as repair of worn-out high valued engineered parts. The functional performance of laser metal deposition process is greatly dependent on its process parameters; therefore, considering the type of job and nature of material, they need to be adequately optimized before a job can be successfully carried out and with the desired properties. The processing parameters that govern the laser metal deposition process include: the laser power, the scanning speed, the powder flow rate and the gas flow rate. A lot of interactions exist among these processing parameters that make the careful optimization of the processing parameters an important task. In this chapter, modelling of laser metal deposition process of metal alloys and composites is presented. The chapter consist of an in depth review of literature on this subject in the introduction (Sect. 1). Optimization of process parameters for laser metal deposition of titanium alloy is presented in Sect. 2. A case study on statistical modelling of titanium alloy composite and process parameters optimization is presented in Sect. 3. The chapter ends with the summary in Sect. 4.

Keywords Additive manufacturing · Laser metal deposition · Modelling · Optimization · Titanium · Wear

1 Introduction

Additive manufacturing method has come to revolutionized the way product are being manufactured and has offer solutions to a number of industrial problems, including reduction of waste in product remanufacturing [1]. Products can be manufactured

R. M. Mahamood (✉) · E. T. Akinlabi
Department of Mechanical Engineering Science, University of Johannesburg, Johannesburg, South Africa
e-mail: mahamoodmr2009@gmail.com

R. M. Mahamood
Department of Mechanical Engineering, University of Ilorin, Ilorin, Nigeria

K. Gupta and M. K. Gupta (eds.), *Optimization of Manufacturing Processes*, Springer Series in Advanced Manufacturing, https://doi.org/10.1007/978-3-030-19638-7_4

with desired material properties using additive manufacturing process through the flexibilities offered by the process having the ability to handle multilateral [2–4]. Additive manufacturing technology can be broadly classified into two, namely: laser based additive manufacturing and non-laser based additive manufacturing technologies [5]. Laser based additive manufacturing technology is the most popular additive manufacturing technology due to the unique properties provided by the laser beam which is the source of energy used in this manufacturing process. The ability to direct laser beam only to the needed area without interfering with the surrounding material. Laser metal deposition process, selective laser sintering and selective laser melting are important laser additive manufacturing technologies. These important technologies are still evolving because, a lot still need to be understood as regards to the physics of these processes. A number of research works has been recorded in the literature towards understanding these processes. Some research works are dedicated to model the process. There is no doubt that modeling and simulation will go a long way to properly understand this process, as against the commonly trial and error methods usually adopted in process optimization when components are built. It will also help to quantify the properties of the manufactured components based on the process parameters employed. The flexibility offered by additive manufacturing process in its ability to produce components with combination of materials makes modelling of these processes more challenging. In order to adequately model these additive manufacturing processes, multiple scales modeling techniques are required. This is to be able to account for the detail properties of the new class of materials being processed through this important manufacturing process. For example, the traditional alloying place limit on the quantity of alloying element that can be permissible by thermodynamic laws. Additive manufacturing has transient this laws and it makes it possible to produce alloys of materials with wide apart melting temperatures. This type of material cannot be alloyed together using the traditional melting process because the two materials will segregate upon cooling. The rapid solidification of the melt pool that characterized laser additive manufacturing technologies makes it difficult for materials to segregate and the melting takes place bit by bit and layer by layer. The use of multiple length scales modelling will help in the development of the basic understanding of the underlying physics of the process and within the macro-scale models to effectively simulate component performance.

Against this background, Francois et al. [6] review some of the challenges facing the modelling and simulation of additive manufacturing processes. This study showed that modeling the process, microstructure, and other properties, and process optimization needs to be done at different scales and regimes, which will enable the advancement towards an integrated computational method in achieving the process-structure-properties-performance which will help optimization of materials to meet a specific performance requirement. This is shown schematically in Fig. 1. Criales et al. [7] investigated the effect of process parameters on temperature profile and melt pool geometry and developed a predictive model for selective laser melting of Inconel 625. The influence of powder parking density on the characteristics of the product was analyzed. The result of this study showed that the proposed finite element model is in good agreement with the experimental data. The temperature distribution with

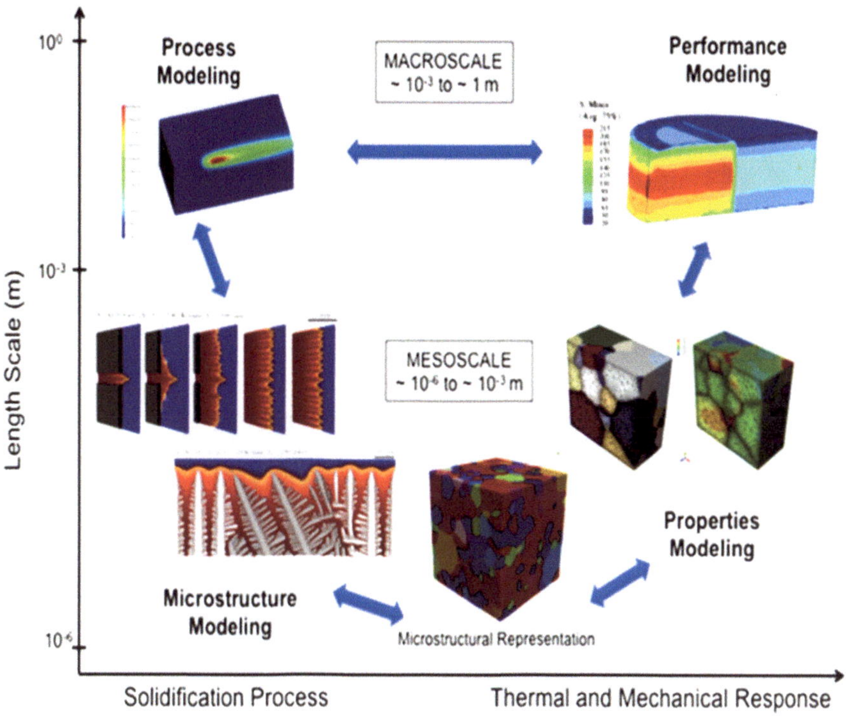

Fig. 1 Illustration of the envisioned integrated process-structure-properties-performance modeling and simulation approach and associated length scales [6]

the melt pool geometry predicted by the model is very close to the melt pool from an experiment conducted by Yadroitsev et al. [8] as shown in Fig. 2. Verma and Rai [9] formulated an optimal process planning model for selective laser sintering process to address the sustainability issue in additive manufacturing. Acharya and Das [10] investigated the modelling of IN100, a high gamma-prime nickel-based superalloy, using scanning laser epitaxy. The result of this study showed that a crack free deposition of IN100 can be produced using the developed model. Comprehensive modelling approaches in additive manufacturing was done by Bikas et al. [11] for the purpose of identifying research gaps and highlighted the importance of modelling in development of closed-loop control for the additive manufacturing process.

Stender et al. [12] proposed a finite element (FE) analysis to simulate the thermal and mechanical responses in laser engineered net shaping (LENS), a laser additive manufacturing process, at a macroscopic length scale to predict thermal conditions during the manufacturing process, distortions, strength and residual stresses. The proposed model allows surfaces for radiation and convection, the coupling of displacements between thermal and mechanical calculations, and explicit consideration of solid and fluid phases to be accurately defined. The results were compared to experimental data of temperature profiles, Vickers hardness mapping, and electron

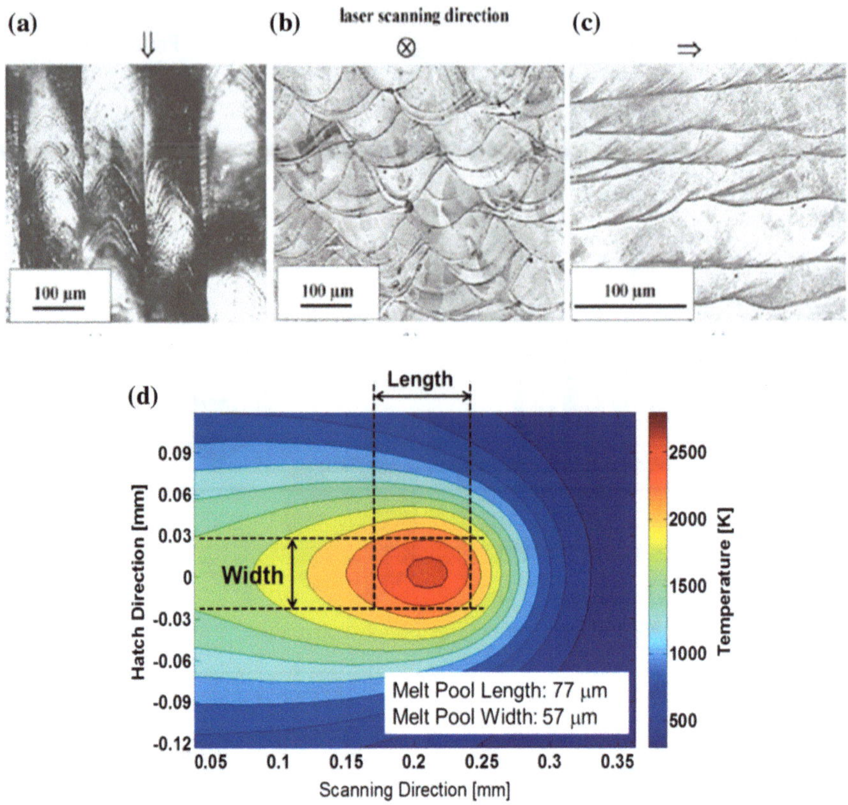

Fig. 2 Measured and predicted melt pool geometry in SLM of Inconel 625: **a** measured melt pool marks on the surface, **b** cross-section, **c** longitudinal section [8], and **d** predicted temperature [7]

backscatter diffraction. There was good agreement between the experiment and modelling results. The modelling approach was able to predict thermal and mechanical responses in the manufacturing process which can be measured in real materials and can be used in design and optimization of the LENS process. Tapia et al. [13] also proposed a predictive model for predicting the porosity in metallic parts fabricated using selective laser melting, also a laser additive manufacturing process. The authors developed a spatial Gaussian process regression model for part porosity as a function process parameters. Bayesian inference framework was then used to estimate the statistical model parameters. Kriging method was employed to predict the part porosity at any given setting.

The model was validated through experiment for predicting the porosity in 17-4 PH stainless steel manufactured using a ProX 100 selective laser melting machine. The algorithm of the proposed model is shown in Fig. 3. The study showed that the proposed model was able to accurately predict porosity in the manufacture samples and was able to reduce porosity to as low as 0.325% at with predicted processing

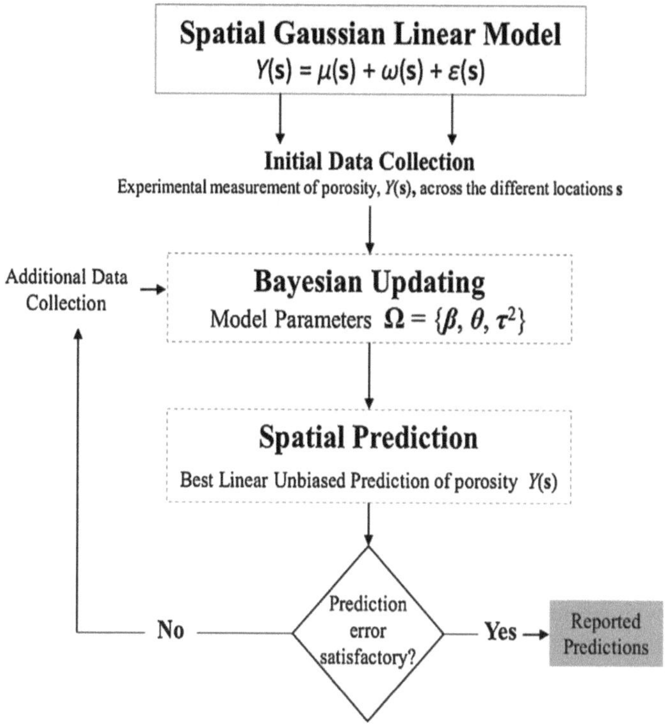

Fig. 3 Summary of the predictive methodology [13]

parameters; a laser power of 50 W and scanning velocity of 275 mm/s. Owing to the importance of modelling and simulation in process development and enhancements, a number of research works has appeared in the literature on modelling the additive manufacturing processes. These effort will ultimately help to further understand these manufacturing processes and also help in the much needed processes maturity that will expand the application of these manufacturing process to the fabrication of more critical parts especially in the aerospace industry. This chapter presents some of these important research works in modelling and processes parameter optimization of laser additive manufacturing technologies. A case study is also presented on process parameter optimization of titanium alloy composite produced with laser metal deposition process, an additive manufacturing technology, using response surface method. The chapter ends with summary.

2 Process Parameter Modelling and Optimization of Laser Additive Manufacturing

Laser additive manufacturing is a promising manufacturing process that is capable of solving engineering problems and also for building innovative engineering machine components. One of the road blocks in the rapid development of this exciting manufacturing process is the ability to adequately predict the material properties resulting from these processes. A lot of research works has been conducted through experimental study to understand the processing parameter influence on the resulting material properties. Experiments are expensive and time consuming which is why model development will go a long way in providing a better understanding of the process and through simulations the process can be effectively controlled. In this section, some of the research works in the literature on modelling and process parameters optimization of laser additive manufacturing technologies are presented. Lee and Prabhu [14] developed a process control models for direct energy deposition and powder bed fusion processes. The models were built using regression metamodel of heat transfer and thermal model that helps to account for residual heat in track-to-track interactions. The two models are coupled using temperatures that were predicted with the auxiliary model as initial conditions for predictions in metamodel for future laser scans. The models are used to generate training data for a model-free optimal controller. The result of the simulation showed that the proposed optimal controller is capable of adjusting the laser scan speed in order to control temperature. Foteinopoulos et al. [15] carried out a similar study by proposing a finite difference (FD) model of the thermal history of parts produced using powder bed fusion Additive Manufacturing (AM). The model was developed by calculating the temperature of the part at each time-step to describe the material thermal properties by taking into account, the moving laser heat source, the melting phase change and the functions of both temperature and porosity. The developed model is capable of simulating and storing the temperature history of a thin walled 3D part using an adaptive meshing strategy that helps to decrease the computational load for the simulation and optimization of the process parameters to improve the production efficiency based on time and energy.

Conti et al. [16] studied a finite element modeling of laser additive manufacturing to predict material performance from the process. The authors have tried to simplify the specific heat dependence and thermal conductivity from temperature to develop the model for assessing the process parameters. Baturynskaa et al. [17] also used finite element method and machine learning techniques to evaluate and optimize additive manufacturing process parameters. Xiao et al. [18] developed a Gibson-Ashby model to predict and optimize the topology, performance and porosity in three types of three dimensional structures namely: Face Centre Cube (FCC), Vertex Cube (VC), and Edge Centre Cube (ECC) structures produced through selective laser melting process. The results revealed that the performance of topological optimization of selective laser melted built lattice structures made from 316L stainless steel is superior and proved that the developed topological optimization technique can be used to build lightweight lattice structure units. Bonada et al. [19] used finite element method to develop an optimisation procedure to control additive manufacturing process and to also increase manufacturing accuracy of printed part. The optimisation procedure was used to increase the Z accuracy in spatial location. The developed model was able to account for the photo curing material parameters and can easily be adapted to different photosensitive materials. It also allows the photo conversion ratio of any spatial location of printed parts to be controlled which makes it possible to define the desired conversion ratio with the manufacturing direction, so as to obtain a more uniform conversion profile. The model was validated with experimental results which show a good agreement with one another.

Zinovieva et al. [20] developed a three-dimensional numerical model to evaluate evolution of grain structure for additive manufacturing. The model was developed using cellular automata and finite difference methods to predict the evolving grain structure based on the transient temperature in selective laser melting process. The developed model was able to represent the columnar grain structure very well which is attributed to characteristic solidification conditions in selective laser melting and associated grain growth that promotes the development of coarse columnar grains with resulting morphological and crystallographic texture as shown in Fig. 4. The model result is in good agreement with the experimental results as shown in Fig. 5. Modelling of additive manufacturing process chains was conducted by Thompson et al. [21] using new two-dimensional approach to modeling manufacturing process chains. In this approach, the role of additive manufacturing technologies in process chains for a part with micro scale features and no internal geometry was used to develop the model. The result of this study showed that additive manufacturing can compete very well with traditional process chains for any small production run. The model can serve as an important tool in manufacturing process selection, concurrent engineering, and design for manufacturing in the early phases of conceptual and detailed design. The efforts of researchers in the area of process modelling in laser additive manufacturing can be seen to be of importance in advancing the technology further as it reduces the financial burden of intensive experimentation. However, more is still needed to be done in this area so that adequate models can be developed that will be capable of predicting and controlling the process more effectively.

Laser metal deposition process for example is an important laser additive manufacturing technology that offers a number of solutions to many manufacturing industries' problems such as the repair of high valued parts and the ability to extend material service life through product remanufacturing. This additive manufacturing technology is very sensitive to the processing parameters and any slight change in the process parameter settings can result in considerable change in the resulting properties. This process need to be adequately modeled so that the process can be adequately controlled using proper close loop control system.

A case study on laser additive manufacturing process is presented in the next section. The case study is based on laser metal deposition process, an important laser additive manufacturing technology [22]. Response surface method of design of experiment was used in a previous study as detailed in a patent filed in 2016 [23]

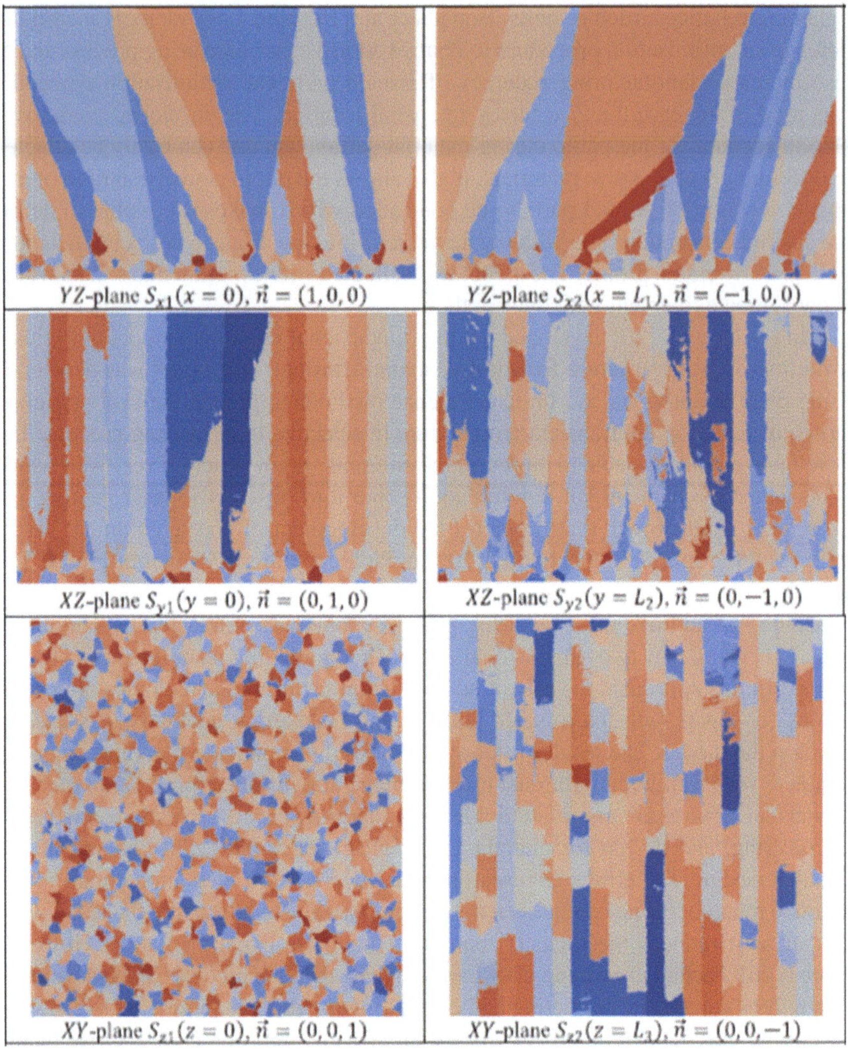

Fig. 4 Grain structures in different boundary planes of the computational volume. Laser beam moved along the Y-axis, from 0 to L2, n! denotes the unit normal vector of a plane [20]

to develop an empirical model. The model was used in this case study for process parameter optimization.

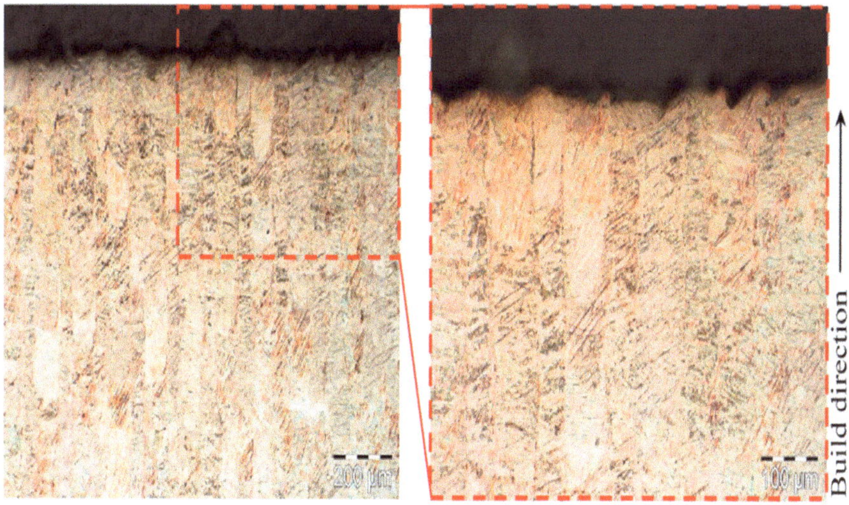

Fig. 5 Experimental grain structure of Ti-6Al-4V alloy subjected to SLM. The scanning strategy was unidirectional, the scanning speed was 1200 mm/s [20]

3 Process Parameters Optimization of Laser Metal Deposited Ti6Al4V-TiC Composite Using Response Surface Method (RSM): A Case Study

Empirical modelling is used to describe models that are constructed from experimental data. In empirical Modelling, the observables or experimental data responses are variables that can take different values and the current values are used to determine the current state of the observables. It may become complicated when the value of one observable affects the output of another observable which is known as interaction. To be able to capture this behaviour in empirical modelling, the experimental data needs to be collected based on experimental design. A careful use of design of experiment (DOE) technique plays a major role to capture the needed causal and effect relationship as well as interactions among the experimental factors. A number of software is now available for the design of experiment and model development. Examples of such software include Minitab and Design Expert. With these software, the experiment can be designed, implemented, analysed and the required model developed. In this case study, the experimental data was collected based on experimental design matrix prepared based on response surface methodology of design of experiment, applied via Design Expert software. The experiments were conducted following the experimental design and the results were uploaded the Design Expert software for analysis and model development. The empirical model used in this case study for process parameter optimization was referenced from [23]. An initial screening experiment was conducted using full factorial design of experiment in order to establish the least significant process parameter from four process parameters. This was required to reduce the number of experiment for the response surface method. The screened process parameters are the laser power, scanning speed, powder flow rate and gas flow rate. The gas flow rate was found to be the least significant process parameter and it was removed from the response surface method. The process parameters used in the response surface method are the laser power, powder flow rate, scanning speed and TiC percentage. Design expert 8 software was used for experimental design, modelling and optimization. Statistical technique Desirability Analysis has been used for optimization. This optimization technique was first presented by Derringer and Suich [24]. In this, the initial step is to change the single response yi (x) into an individual desirability function (di) and differ over the extent $0 \leq di \leq 1$. The desirability function can be divided into three categories on the basis of kind of response characteristics:

1. "Higher is better",

$$di = \begin{cases} 0, & y_i \leq y_{i*} \\ \left[\frac{y_i - y_{i*}}{y_i^{'} - y_{i*}}\right]^t, & y_{i*} < y_i < y_i^{'} \\ 1, & y_i \geq y_i^{'}, \end{cases} \tag{1}$$

where y_{i*} is the minimum adequate value of y_i, $y_i^{'}$ is the maximum value of y_i and t describes the shape function for desirability.

2. "Smaller is better",

$$di = \begin{cases} 1, & y_i \leq y_i^{''} \\ \left[\frac{y_i^* - y_i}{y_i^* - y_i^{''}}\right]^r, & y_i^{''} < y_i < y_i^* \\ 0, & y_i \geq y_i^*, \end{cases} \tag{2}$$

where $y_i^{''}$ is the minimum value of y_i, y_i^* is the highest adequate value of y_i and r describes the shape function.

3. "Nominal is better",

$$di = \begin{cases} \left[\frac{y_i - y_i^*}{C_i - y_i^*}\right]^s, & y_{i*} < y_i < C_i \\ \left[\frac{y_i - y_i^*}{C_i - y_i^*}\right]^t, & C_i < y_i < y_i^* \\ 0, & y_i > y_i^* \text{ or } y_{i*} > y_i \end{cases} \tag{3}$$

where C_i is the mainly adequate or objective value, s and t describes the exponential parameters which verify the shape of desirability function.

Overall desirability function of the multi-response is presented as $D = (d_i^{w1} . d_2^{w2} \ldots d_n^{wn})$, where w_j ($0 < w_j < 1$) is the weight value given for the importance of jth response variable and $\sum_{j=1}^{n} w_j = 1$. The combination of parameters with highest desirability is considered as the optimum factor.

Table 1 Constraints to minimize the wear volume loss

Parameter	Goal	Lower limit	Upper limit
A: laser power	is in range	1.6	2.8
B: scanning speed	is in range	0.005	0.01
C: powder flow rate	is in range	2	4
D: vol% TiC	is in range	20	50
Wear volume	Minimize	0.009744	0.0886989

The optimization in this case study was achieved by setting the constraints and setting the goal for optimization, which is 'smaller-is-better', for minimizing the wear volume, (that maximizes the wear resistance property of the composite). The optimization was done by setting the goal for wear-volume optimization. After entering the constraints into the software, the software was allowed to search for optimized process parameters that resulted in minimum wear volume loss. The summary of the constrained entered into the model is presented in Table 1.

The model used the constraint parameters presented in Table 1 to search for the sets of optimised processing parameters that would minimise the wear volume. A total of 47 solutions were found, as shown in Table 2. The software also suggested one of the results by underlining selected in the remark column of the table as seen in Table 2. From Table 2, it can be seen that the optimisation for a wear volume as low as 0.009, is achievable. The suggested optimization result is taken and analyzed. The surface plot of the optimised setting for the selected result is shown in Fig. 6.

Table 2 Solutions for the optimization problem

Number	Laser power	Scanning speed	Powder flow rate	vol% TiC	Wear volume	Desirability	Remark
1	2.08	0.0083	3.00	43.66	0.009	1.000	Selected
2	2.17	0.0088	3.08	43.35	0.010	1.000	
3	2.17	0.0078	2.89	42.18	0.010	1.000	
4	2.03	0.0081	2.96	43.31	0.010	1.000	
5	2.24	0.0081	3.00	41.86	0.010	1.000	
6	2.25	0.0083	3.21	41.97	0.010	1.000	
7	2.15	0.0080	3.00	41.28	0.010	1.000	
8	2.16	0.0085	3.20	44.03	0.009	1.000	
9	2.14	0.0075	3.07	42.74	0.010	1.000	
10	2.15	0.0087	3.11	43.41	0.009	1.000	
11	2.14	0.0080	2.94	41.09	0.010	1.000	
12	2.16	0.0081	3.22	44.36	0.009	1.000	
13	2.26	0.0082	3.16	42.08	0.009	1.000	

(continued)

Table 2 (continued)

Number	Laser power	Scanning speed	Powder flow rate	vol% TiC	Wear volume	Desirability	Remark
14	2.08	0.0084	3.07	43.62	0.009	1.000	
15	2.17	0.0081	3.05	42.13	0.009	1.000	
16	2.20	0.0079	2.90	42.91	0.010	1.000	
17	2.13	0.0084	3.01	41.72	0.010	1.000	
18	2.31	0.0083	3.23	45.99	0.010	1.000	
19	2.04	0.0081	3.10	45.15	0.009	1.000	
20	2.11	0.0082	3.15	48.15	0.009	1.000	
21	2.12	0.0085	3.05	48.47	0.009	1.000	
22	2.19	0.0086	3.10	41.85	0.010	1.000	
23	2.28	0.0084	3.07	46.10	0.009	1.000	
24	2.26	0.0088	3.15	43.40	0.010	1.000	
25	2.21	0.0077	3.18	44.70	0.010	1.000	
26	2.06	0.0084	3.02	46.34	0.009	1.000	
27	2.13	0.0081	3.16	47.17	0.009	1.000	
28	2.02	0.0085	2.96	45.89	0.010	1.000	
29	2.20	0.0087	3.23	46.06	0.009	1.000	
30	2.11	0.0080	3.17	42.52	0.009	1.000	
31	2.25	0.0088	3.13	49.25	0.010	1.000	
32	2.22	0.0083	3.16	46.54	0.009	1.000	
33	2.08	0.0081	3.01	44.85	0.009	1.000	
34	2.11	0.0082	3.18	41.78	0.010	1.000	
35	2.29	0.0081	3.22	44.41	0.010	1.000	
36	2.01	0.0083	3.19	46.69	0.010	1.000	
37	2.16	0.0080	3.21	41.80	0.009	1.000	
38	2.17	0.0081	3.26	44.61	0.009	1.000	
39	2.18	0.0081	3.21	43.61	0.009	1.000	
40	2.08	0.0088	3.11	47.55	0.010	1.000	
41	2.24	0.0089	3.16	44.75	0.010	1.000	
42	2.15	0.0085	3.05	47.36	0.009	1.000	
43	2.16	0.0087	3.13	45.93	0.009	1.000	
44	2.17	0.0087	3.37	46.35	0.010	1.000	
45	2.23	0.0084	3.32	47.68	0.010	1.000	
46	2.16	0.0080	3.08	44.26	0.009	1.000	
47	2.18	0.0083	3.16	47.51	0.009	1.000	

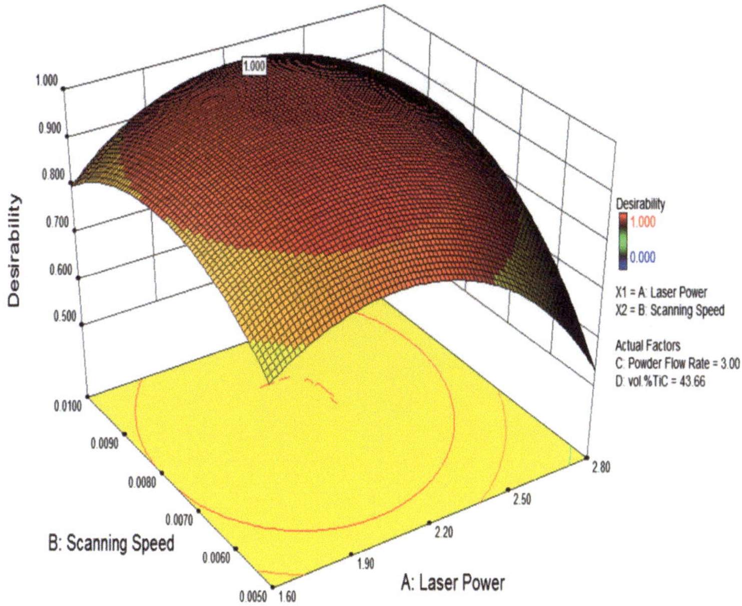

Fig. 6 Surface plot for the optimized setting showing the desirability versus scanning speed and laser power

Another optimization was performed by fixing the TiC percentage to 20%, and setting the optimization goal to 'minimize the wear volume'. The laser power, the scanning speed and the powder flow rate were kept in range in order for that the model to search for the appropriate laser power, scanning speed and powder flow rate values that would minimize the wear-volume loss and within the given range. In the first optimization process, the percentage TiC was kept in range of between 20 and 50. And the optimized percentage TiC was found to be 43.66%. This current optimization process finds the optimized process parameters while keeping the percentage TiC to 20% and at the same time minimizing the wear volume loss. Table 3 presents the summary of the optimization constraints that was fed into the model.

Table 3 Optimization constraints for the 20 vol% TiC

Name	Goal	Lower limit	Upper limit
A: laser Power	is in range	1.6	2.8
B: scanning speed	is in range	0.005	0.01
C: powder flow rate	is in range	2	4
D: vol% TiC	is equal to 20.00	20	50
Wear volume	Minimize	0.009744	0.0886989

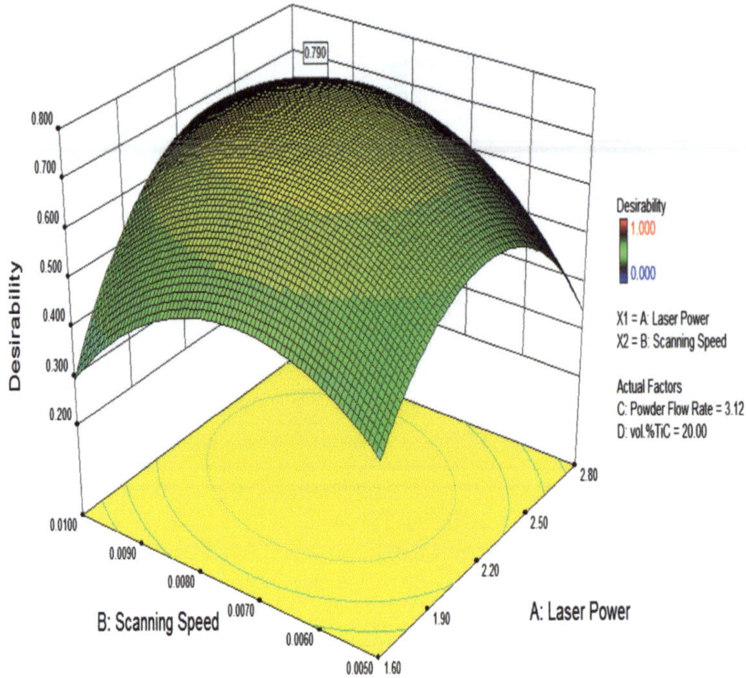

Fig. 7 Surface plot for the optimized process parameters at 20% TiC

The software found only one solution, which is presented as follows:

Laser Power: 2.27 kW
Scanning Speed: 0.0078 m/s
Powder Flow Rate: 3.12 g/min
vol% TiC: 20
Wear Volume: 0.026 mm^3
Desirability: 0.790.

The surface plot for the desirability versus the processing parameters is shown in Fig. 7.

The third optimization problem was setting the laser power, the scanning speed, the powder flow rate, and %TiC to the fixed values, and using the model and in the Design Expert 8 software environment to predict the wear volume loss. The constraints are presented in Table 4.

Only one solution was found that is as follows:

Laser Power: 2.0 kW
Scanning Speed: 0.006 m/s
Powder Flow Rate: 2.5 rpm
vol% TiC: 50%
Wear Volume: 0.018 mm^3.

Table 4 The constraints with fixed process parameters to predict wear-volume

Name	Goal	Limit	Limit
A: laser power	is equal to 2.00	1.6	2.8
B: scanning speed	is equal to 0.0060	0.005	0.01
C: powder flow rate	is equal to 2.50	2	4
D: vol% TiC	is equal to 50.00	20	50
Wear volume	is in range	0.009744	0.0886989

Table 5 Validation of optimization results

Number	Laser power	Scanning speed	Powder flow rate (rpm)	vol% TiC	Predicted wear volume	Actual wear volume
1	2.08	0.0083	3.00	43.66	0.009	0.012
2	2.27	0.0078	3.12	20	0.026	0.027
3	2.0	0.006	2.5	50	0.018	0.022

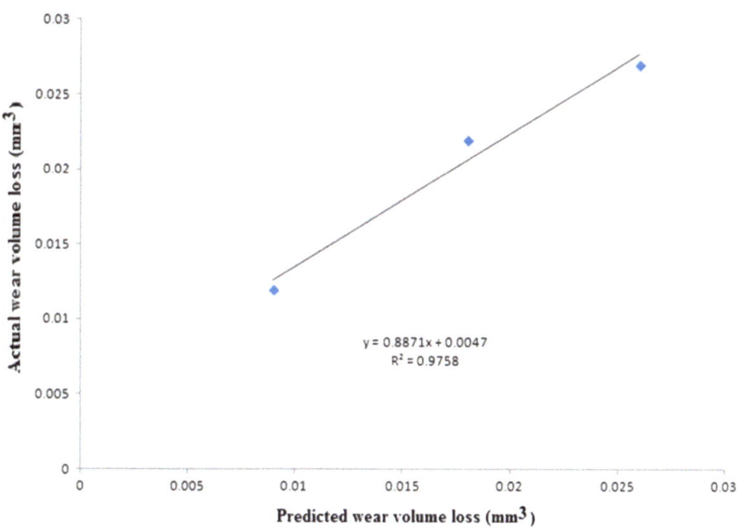

Fig. 8 Graph of actual wear volume loss against the predicted optimized wear loss

In order to validate these process parameters optimization results, experiments were conducted using the process parameters in the optimization results. These experiments were repeated twice; and the wear tests were conducted on the samples. The average wear volume loss for each set of the processing parameters are reported in Table 5.

In order to compare the experimental results with the predicted optimized process parameters that minimizes wear volume, a graph of the experimental or actual wear volume loss against the predicted wear volume loss results is shown in Fig. 8.

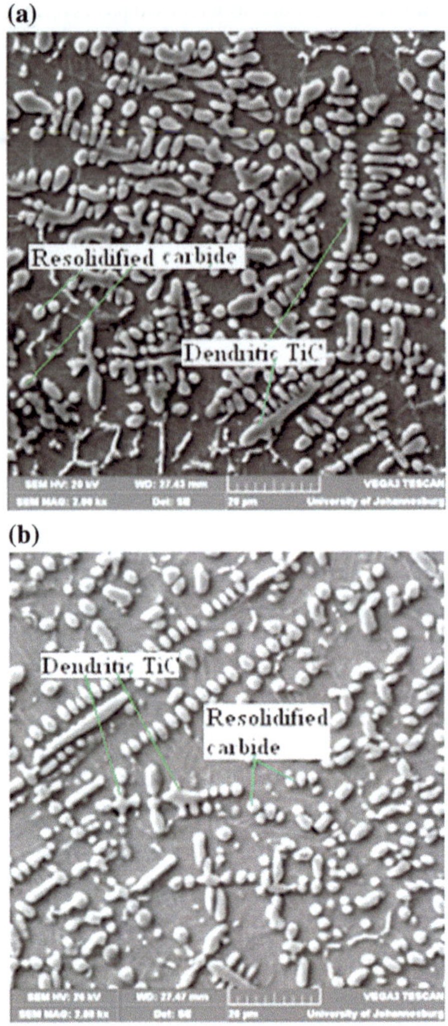

Fig. 9 The SEM micrograph of sample at: **a** laser power of 2.08 kW, scanning speed of 0.0083 m/s, powder flow rate of 3 rpm and % TiC of 43.66, **b** laser power of 2.0 kW, scanning speed of 0.006 m/s, powder flow rate of 2.5 rpm and % TiC of 50

The graph shows that the actual experimental values are close to the predicted values, and the slight differences could be attributed to some experimental uncertainties, which are still within a tolerable limit. The microstructures of the samples 1 and 3 are presented in Fig. 9 and the wear tracks of the two samples are also shown in Fig. 10. The microstucture of the sample number 1 is characterized by large quantity of dendritic TiC and fewer resolidified carbide are also seen as indicated in Fig. 9. The microstructure of the sample 3 on the other hand consists of larger quantity of resolidified carbide and fewer dendritic TiC. The better wear resistance performance of sample 1 can be attributed to the less resolidified carbide and the larger quantity of dendritic TiC which is harder and become grinded into powder during the slidding

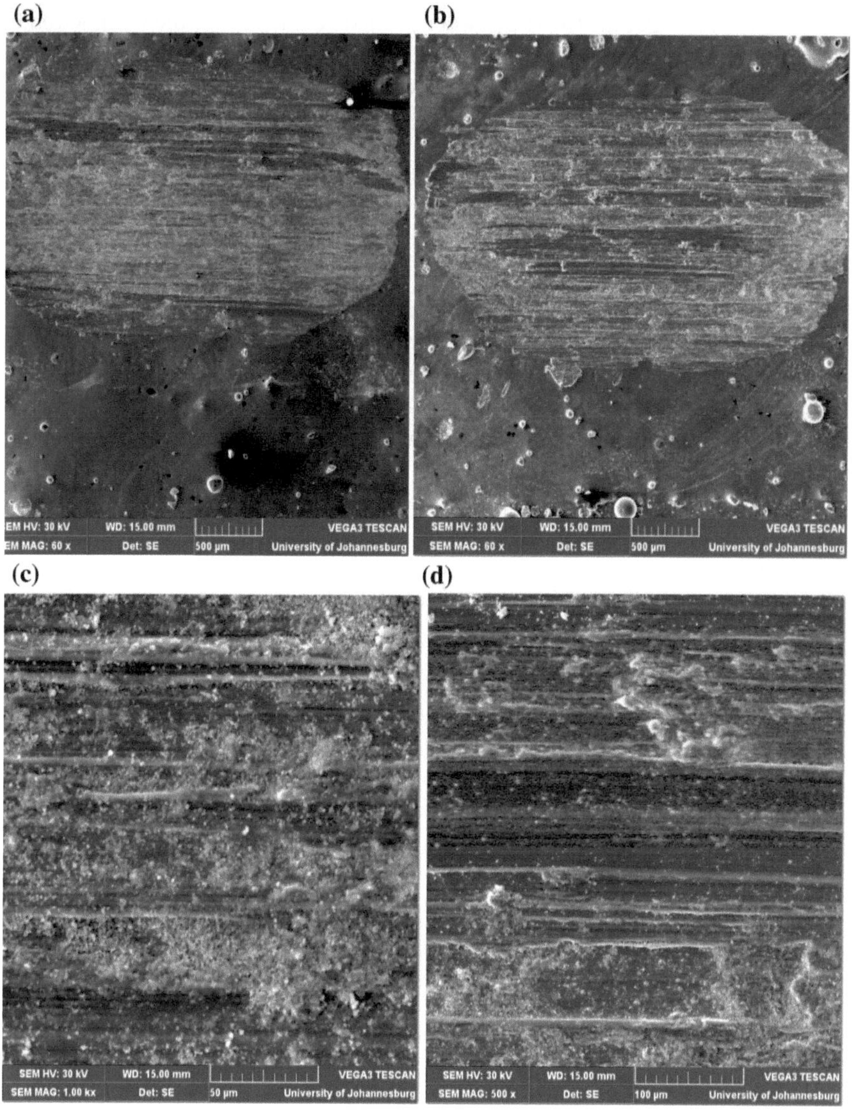

Fig. 10 The SEM micrograph for the wear track of sample at: **a** laser power of 2.08 kW, scanning speed of 0.0083 m/s, powder flow rate of 3 rpm and % TiC of 43.66, **b** laser power of 2.0 kW, scanning speed of 0.006 m/s, powder flow rate of 2.5 rpm and % TiC of 50, **c** higher magnification of (**a**), **d** higher magnification of (**b**)

wear test as shown in Fig. 10. The powder formed help to reduce the wear action by forming a power lubricant. The sample 3 also behave in the same manner but not as better as that of sample 1 because of fewer quantity of dendritic TiC as compared to the larger ones seen in sample 1.

4 Summary

This chapter has summarized modelling and optimization of laser additive manufacturing process. A case study in process parameter optimization using response surface method and desirability technique is also presented. The results showed that the model was able to predict optimize processing parameters and are in good agreement with the experimental data. It is concluded that modeling and optimization is a key to understand the process further, establish the relationship between parameters and outputs, and improve the process performance. Enough attempts have been made in this area, however a vast scope exists for future work for modeling and optimization of all laser based additive manufacturing processes for repair, maintenance, and remanufacturing applications using various statistical and evolutionary techniques to establish the filed further.

Acknowledgements This work is supported by the Rental Pool Programme of National Laser Centre, Council of Scientific and Industrial Research (CSIR-NLC), Pretoria, South Africa.

References

1. Mahamood RM, Akinlabi ET, Owolabi MG (2017) Laser metal deposition process for product remanufacturing. In: Gupta K (ed) Advanced manufacturing technologies. Springer, Switzerland, pp 267–291
2. Mahamood MR, Akinlabi ET (2017) Functionally graded materials. Springer Science Publisher, Switzerland
3. Mahamood RM, Akinlabi ET (2015) Effect of processing parameters on wear resistance property of laser material deposited titanium-alloy composite. J Optoelectron Adv Mater (JOAM) 17(9–10):1348–1360
4. Mahamood RM, Akinlabi ET (2015) Laser metal deposition of functionally graded Ti6Al4V/TiC. Mater Des 84:402–410
5. Mahamood RM, Akinlabi ET, Shukla M, Pityana S (2014) Revolutionary additive manufacturing: an overview. Lasers Eng 27:161–178
6. Francois MM, Sun A, King WE, Henson NJ, Tourret D, Bronkhorst CA, Carlson NN, Newman CK, Haut T, Bakosi J, Gibbs JW, Livescu V, Vander Wiel SA, Clarke AJ, Schraad MW, Blacker T, Lim H, Rodgers T, Owen S, Abdeljawad F, Madison J, Anderson AT, Fattebert J-L, Ferencz RM, Hodge NE, Khairallah SA, Walton O (2017) Modeling of additive manufacturing processes for metals: challenges and opportunities. Curr Opin Solid State Mater Sci 21:198–206
7. Criales LE, Arısoy YM, Özel1 T (2016) Sensitivity analysis of material and process parameters in finite element modeling of selective laser melting of Inconel 625. Int J Adv Manuf Technol 86:2653–2666

8. Yadroitsev I, Thivillon L, Betrand P, Smurov I (2007) Strategy of manufacturing components with designed internal structure by selective laser melting of metallic powder. Appl Surf Sci 254:980–983
9. Verma A, Rai R (2017) Sustainability-induced dual-level optimization of additive manufacturing process. Int J Adv Manuf Technol 88:1945–1959
10. Acharya R, Das S (2015) Additive manufacturing of IN100 superalloy through scanning laser epitaxy for turbine engine hot-section component repair: process development, modeling, microstructural characterization, and process control. Metall Mater Trans A 46a:3864–3875
11. Bikas H, Stavropoulos P, Chryssolouris G (2016) Additive manufacturing methods and modelling approaches: a critical review. Int J Adv Manuf Technol 83:389–405
12. Stender ME, Beghini LL, Sugar JD, Veilleux MG, Subia SR, Smith TR, San Marchi CW, Brown AA, Dagel DJ (2018) A thermal-mechanical finite element workflow for directed energy deposition additive manufacturing process modelling. Addit Manuf. https://doi.org/10.1016/j.addma.2018.04.012
13. Tapia G, Elwany AH, Sang H (2016) Prediction of porosity in metal-based additive manufacturing using spatial Gaussian process models. Addit Manuf 12:282–290
14. Lee J, Prabhu V (2016) Simulation modeling for optimal control of additive manufacturing processes. Addit Manuf 12:197–203
15. Foteinopoulos P, Papacharalampopoulos A, Stavropoulos P (2018) On thermal modeling of additive manufacturing processes. CIRP J Manuf Sci Technol 20:66–83
16. Conti P, Cianetti F, Pilerci P (2018) Parametric finite element model of SLM additive manufacturing process. Procedia Struct Integrity 8:410–421
17. Baturynskaa I, Semeniutaa O, Martinsena K (2018) Optimization of process parameters for powder bed fusion additive manufacturing by combination of machine learning and finite element method: a conceptual framework. Procedia CIRP 67:227–232
18. Xiao Z, Yang Y, Xiao R, Bai Y, Song C, Wang Di (2018) Evaluation of topology-optimized lattice structures manufactured via selective laser melting. Mater Des 143:27–37
19. Bonada J, Muguruza A, Fernández-Francos X, Ramis X (2018) Optimisation procedure for additive manufacturing processes basedon mask image projection to improve Z accuracy and resolution. J Manuf Processes 31:689–702
20. Zinovieva O, Zinoviev A, Ploshikhin V (2018) Three-dimensional modeling of the microstructure evolution during metal additive manufacturing. Comput Mater Sci 141:207–220
21. Thompson MK, Stolf A, Mischkot M (2016) Process chain modeling and selection in an additive manufacturing context. CIRP J Manuf Sci Technol 12:25–34
22. Mahamood MR (2018) Laser metal deposition process of metals, alloys, and composite materials. Springer, Switzerland
23. Mahamood RM, Akinlabi ET, Shukla M, Pityana S (27 July 2016) Process for the manufacture of a titanium composite using additive manufacturing. Application no. 2016/04998. Patent J 49(7) (Part 2, 2):78
24. Derringer G, Suich R (1980) Simultaneous optimization of several response variables. J Qual Technol 12:214–219

Prediction and Optimization of Tensile Strength in FDM Based 3D Printing Using ANFIS

Shilpesh R. Rajpurohit and Harshit K. Dave

Abstract Fused Deposition Modeling (FDM) is universally used 3D printing technology, to manufacture prototypes as well functional parts due to its capability to create components having any geometric complexity in shorter duration, without any specific tooling requirement or human intervention. FDM fabricated parts have found many promising application in various industries such as aerospace, automobile, medical, customizable products etc. However, the application of FDM parts has been restricted by poor mechanical performance. The mechanical properties of the FDM fabricated part are largely affected by selection of various build parameters. Optimal selection of various build parameters can help to achieve better mechanical strength. The Adaptive network-based a fuzzy Interference System (ANFIS) is uses both neural networks and fuzzy logic to generate a mapping between inputs and response. In ANFIS, the parameters for fuzzy system has been identifying using a neural network. Hybrid learning rule can be used for creating a fuzzy set of IF-THEN rules with the appropriate membership functions and generating previously defined Input/Outputs pairs. Initially, a detailed experimental investigation was conducted to understand the impact of different build parameters on the tensile strength of printed PLA. Using experimental data, an optimized model of ANFIS was developed to anticipate the tensile strength of printed parts.

Keywords Fused deposition modeling · Tensile strength · ANFIS · Fuzzy logic · Membership function

1 Introduction

3D printing techniques have tremendous application in the various engineering field, due to a reduction in product development time and able to make a part with any complex geometry [1, 2]. FDM is widely used AM techniques owing to its simplicity,

S. R. Rajpurohit · H. K. Dave (✉)
Department of Mechanical Engineering, S V National Institute of Technology, Surat 395 007, Gujarat, India
e-mail: harshitkumar@yahoo.com

K. Gupta and M. K. Gupta (eds.), *Optimization of Manufacturing Processes*, Springer Series in Advanced Manufacturing, https://doi.org/10.1007/978-3-030-19638-7_5

inexpensive and ease in operation. Fused deposition modelling (FDM) is widely used for creating three-dimensional polymer components by deposition of filament material layer by layer through liquefier nozzle with a movement in X-Y plane. The molten material is extruded through heated liquefier wherein filament material is heated to semi-solid deposited through nozzle. After a layer is deposited, build table is moved downward in z-direction and another layer is deposited. Then the whole process is repeated until the whole object is printed [1, 2]. The schematic diagram of the FDM process is illustrated in Fig. 1.

FDM fabricated parts are widely used in many industries like automobile, aerospace, medicine, electronics, and customer product industries etc. However, further use of the FDM produced object is restricted owing to the poor mechanical performance. The mechanical properties of FDM parts strongly depend on the selection of the build parameters. Mechanical performance of the FDM object can be enhanced by the appropriate selection of build parameter during the part fabrication stage [3]. Hence, the proper section of build parameter and the modeling of the FDM build parameter gaining attentions among the research community. In the view of this, many researchers have tried to model and formulate the relevance between the FDM build parameter and output to enhance the performance of the FDM processed part.

Papazetis and Vosniakos [4] have used ANN to forecast the shape fidelity and material extrusion in fused deposition modelling process to retain and obtain defect free parts. Alimardani and Toyserkani [5] has developed an ANFIS based model to estimate the clad height as a function of with respect to build parameter for laser based additive manufacturing process. They developed a model with 0.07% absolute error to predict the clad height. Garg et al. [6] have used M5 genetic programming

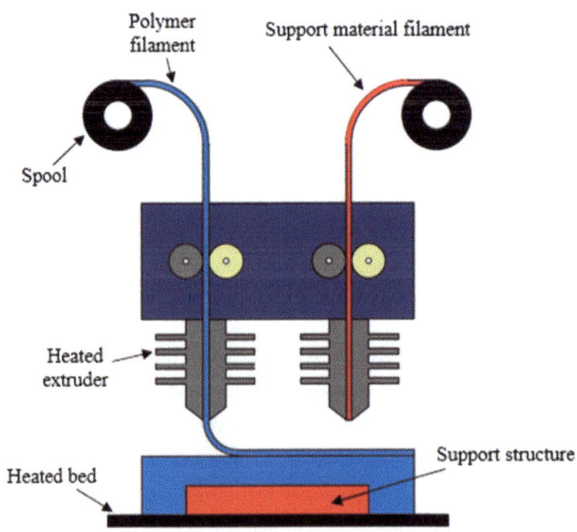

Fig. 1 Schematic representation of fused deposition modeling process

to develop empirical modelling for FDM process variables with an aim to predict the compressive strength. Further, they compared the proposed model with support vector regression (SVR) and ANFIS model. They observed better performance of the proposed model compared to SVR and ANFIS model. Mohamed et al. [7] used a definitive screening approach and ANN to assess the effect of build parameters on creep properties of printed object. Panda et al. [8] used an evolutionary algorithm to formulate the relationship between bead dimensions as response and peak current, travel speed and wire speed as input process parameters. They observed that gene expression programming (GEP) model performed better than multi-gene genetic programming (MGGP). Panda et al. [9] did a comparative study to develop a relationship between tensile strength and build parameters (layer height, raster orientation, bead width and air gap). They found differential evolution algorithm (DEA) perform better than particle swarm optimization (PSO) method. Peng et al. [10] used response surface methodology (RSM) in a combination with fuzzy interface system to optimize the build parameters. Raju et al. [11] used a combination of PSO and bacterial foraging optimization algorithm to optimize the build parameters. Sood et al. [12] used a grey Taguchi method to improve the part accuracy of the FDM processed components. Sood et al. [13] used a hybrid artificial neural network (ANN)—bacterial foraging optimization algorithm (BFOA) to optimize the build process variable to have better mechanical strength. Boschetto et al. [14] used a NN to forecast the surface roughness of the FDM processed samples. Rayegani and Onwubolu [15] used GMDH and differential evolution algorithm to forecast the tensile strength of the fabricated part as a function of build parameters.

It has been observed that some work has been carried out with different types of intelligent techniques to optimize and modelling of FDM build parameters to enhance the quality of the printed part. Mostly various techniques have been used to enhance the surface quality and part accuracy of the FDM printed part. Few works have been reported that discussing enhancement of mechanical quality of printed part using intelligent techniques. However, optimization and modeling of build parameters with ANFIS are also lacking. With the aim of the build parameter optimization such as raster orientation, layer thickness and raster width are selected as a control parameter, input variable and tensile strength is selected as output response, evaluation index. Further, in this study, ANFIS model has been developed to estimate the tensile strength of FDM processed sample.

2 Adaptive Neuro Fuzzy Interface System

ANN and fuzzy logic are combined used by the ANFIS model. Fuzzy logic is a useful tool to convert qualitative approach into crisp output. However, it does not have any defined method for conversion and it is takes a longer time to accommodate the membership functions. While ANN has a greater learning capability to cope-up with the environment. Therefore, fuzzy logic system can use ANN to accommodate fuzzy logic membership functions.

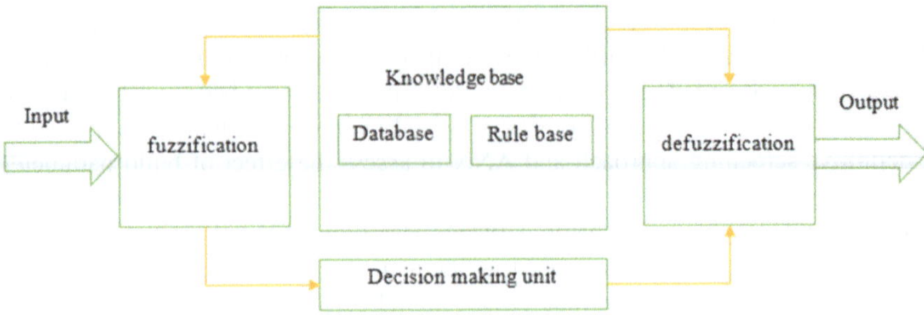

Fig. 2 FIS system

2.1 Fuzzy Interface System

A FIS is designed on three basic components, based on the choice of fuzzy logic rule "If-Then", fuzzy set membership and technique inference. Use fuzzy interface from basic rule to get output. Detailed structure of FIS is shown in Fig. 2. FIS convert the actual value into fuzzy value using the membership functions whose fuzzy values are between 0 and 1. Knowledge base includes basic rules and databases that are two important elements of decision-making. Generally, the database contains definitions such as fuzzy set parameter information with functions defined for existing language variables. Database development typically involves defining the universe, determining the number of language values used for each language variable, and creating membership functions. Depending on the rules, include the fuzzy logic operator and the If-Then conditional statement. The basic rule can be built with or without human intervention and here a search rule using numerically input-output data [10, 16].

2.2 Adaptive Network

The adaptive network is an example of a multilayer feed forward neural network. Figure 3 shows the basic adaptive network. The adaptive network contained a plurality of directly interconnected adaptation nodes, with architectural characteristics without a weight value between them. The error in the output can be reduced by the proper selection of the leaning rules in the adaptive network [16–18].

Basic Adaptive Network learning typically uses a back propagation and spread chain rule or gradient. Since the learning algorithm is an adaptive network, gradient descent or return propagation are always used until date. However, the back propagation algorithm still has weaknesses and can degrade the performance of the adaptive network for decision-making. Slow convergence tends to always stay local minimum, which poses a big problem of back propagation algorithm. Therefore,

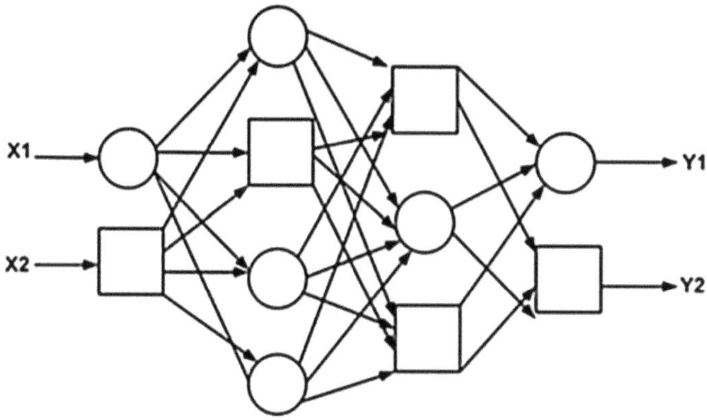

Fig. 3 Adaptive network

a hybrid-learning algorithm was proposed with a better performance to speed up convergence and avoid being trapped by local minimum [16–18].

2.3 ANFIS Architecture

ANFIS is one of the widely used fuzzy interface systems, especially for real physical problems. ANFIS combines both fuzzy logic and ANN, and the neural network method is used to adjust fuzzy interface system parameters. The ANFIS structure consists of five network layers and a hybrid learning algorithm is used to adjust the system according the input and output data structures. An ANFIS type Takagi–Sugeno's schematic diagram with two inputs (x and y) and an output (z), two membership functions for each input and two rules illustrated in Fig. 4.

First order Sugeno FIS is an easy example of a fuzzy interface system. First order Sugeno model consists of two fuzzy rules "If-Then" shown as below:

$$\text{Rule 1}: \text{ If } X \text{ is } A_1 \text{ and } y \text{ is } B_1 \text{ Then } f_1 = p_1x + q_1y + r_1 \tag{1}$$

$$\text{Rule 2}: \text{ If } X \text{ is } A_2 \text{ and } y \text{ is } B_2 \text{ Then } f_2 = p_2x + q_2y + r_2 \tag{2}$$

where, A_1, B_1, A_2 and B_2 are the parameter for the input functions, p_1, q_1, r_1, p_2, q_2 and r_2 are the parameters for output functions.

Each layer of the ANFIS structure is explained as below:

Layer 1: In fuzzification layer, every input node is an adaptive node that converted into linguistics with the use of membership function. The output of the adaptive node can be followed as:

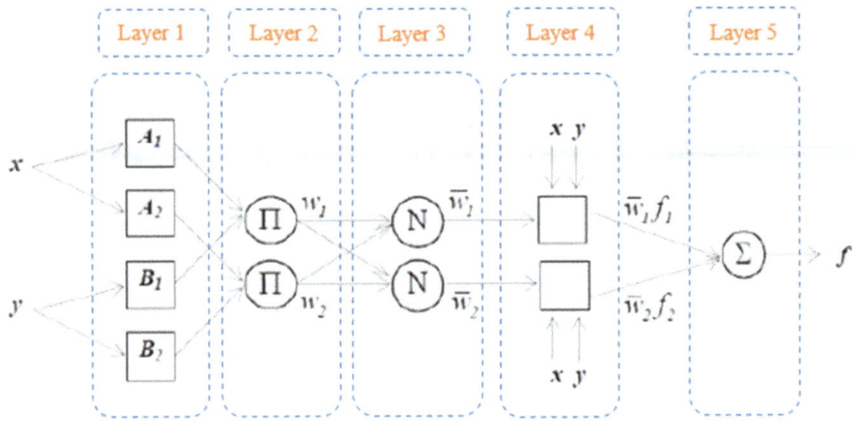

Fig. 4 ANFIS architecture

$$O_{1,i} = \mu_{Ai}(x), \quad i = 1, 2 \tag{3}$$

where μ_{Ai} the degree of membership functions for input x.

Layer 2: In product layer, each node is non-adaptive. Each fixed node in the present layer multiples all the incoming signals and the firing strength of each rule can be calculated as follows.

$$O_{2,i} = \omega_i = \mu_{Ai}(x)\mu_{Bi}(y), \quad i = 1, 2 \tag{4}$$

Layer 3: In the normalization layer, all nodes are fixed. Each node is representing the normalization of firing strength from layer 2. The normalized firing strength can find out as follows.

$$O_{3i} = \bar{\omega}_i = \frac{\omega_i}{\sum_i \omega_i} \tag{5}$$

Layer 4: The de-fuzzification layer is an adaptive layer. The relation between output and input can be expressed as follows.

$$O_{4i} = \bar{\omega}_i f_i = \bar{\omega}_i (\mathrm{p}_i \mathrm{x} + q_i y + r_i) \tag{6}$$

where ω_i is output from the previous layer and $(\mathrm{p}_i \mathrm{x} + q_i y + r_i)$ is a consequent parameter.

Level 5: The output layer represents the modelled output by ANFIS network.

$$O_{5i} = \sum_i \bar{\omega}_i f = \frac{\sum_i \omega_i f}{\sum_i \omega} \tag{7}$$

The training data set contained the input/output pairs needed to form the ANFIS model for predicting the target outcome. The ANFIS model adaptively maps inputs and outputs using different MFs, rule bases, and associated parameters acquired by the loaded learning dataset.

2.4 *Hybrid Learning Algorithm*

The purpose of the learning algorithm in the ANFIS model is to adapt parameters, so that the ANFIS output can be mapped with learning data. The hybrid-learning algorithm uses both gradient descendent and least square method. In the forward passage, least square used to identify the parameters while going in the opposite direction, a gradient descendent method is used. The result of the proposed model can be determined as follows.

$$f = \frac{\omega_1}{\omega_1 + \omega_2} f_1 + \frac{\omega_2}{\omega_1 + \omega_2} f_2 \tag{8}$$

$$f = \bar{\omega}(\mathrm{p}_1 \mathrm{x} + q_1 \mathrm{y} + r_1) + \bar{\omega}(\mathrm{p}_2 \mathrm{x} + q_2 \mathrm{y} + r_2) \tag{9}$$

$$f = (\bar{\omega}_1 x)\mathrm{p}_1 + (\bar{\omega}_1 y)\mathrm{q}_1 + (\bar{\omega}_1)\mathrm{r}_1 + (\bar{\omega}_2 x)\mathrm{p}_2 + (\bar{\omega}_2 y)\mathrm{q}_2 + (\bar{\omega}_2)\mathrm{r}_2 \tag{10}$$

where p_1, q_1, r_1, p_2, q_2 and r_2 are the consequent parameters.

The hybrid learning is very efficient learning algorithm to train the ANFIS model that approach converges much faster compared to back propagation algorithm.

3 Experimental Plan and Procedure

3.1 *Process Parameters and Experimental Plan*

In the present investigation, effect of raster angle, raster width and layer height have been evaluated as a function of tensile strength. The raster angle is defined as the angle of the raster relative to the x-axis of the machine. Five different values of the raster angle are considered and varied as 0°, 30°, 45°, 60° and 90°. The height of the layer can be defined by the height of deposited layer. Five different values of the layer height are considered to be 100, 150, 200, 250 and 300 μm. The raster width can be defined as the width of the bead of material during the deposition of the raster. Four different raster width values are chosen and change to 400, 500, 600 and 700 μm. The layer height and raster width of the selected layer within the possible range of the open source printer being used.

In this study, a full factorial design is used to perform experiments on all combinations of factor levels. Two factors have been varied at the five levels and one factor has been varied at four levels so according to full factorial experimental design, total 100 number of experiments need to be performed. Two identical specimens were constructed for each experiment and a total of 200 specimens were obtained for all the factorial test models.

3.2 Sample Fabrication

In the present investigation, raster angle and layer height have been varied at five levels and raster width has been varied at four levels. After consideration of all possible combination of process parameters, total of 100 experiments have been performed and entire set of experiments have been performed twice. Using PLA filament, all the specimen were built with an OMEGA Dual Extruder, a high precision open source FDM printer. Figure 5 shows a schematic diagram of test components modeled in Pro Engineering and save as STL files. Then STL file was transferred to machine software to create the deposition strategy and to adjust all build parameters. Test specimens were created using constant process parameters as shown in Table 1. The samples were made with the same brand of PLA filament coils and to retain the same properties of the filamentary material. Figure 6 shows the FDM machine used in the present investigation to manufacture test specimen.

3.3 Tensile Testing

To evaluate tensile properties, tensile test has been performed by using Tinus Olsen H50KL universal testing machine. The machine has a load cell of 50 kN and a built-in Horizon software. The testing speed is maintained at 5 mm/min for tensile specimens

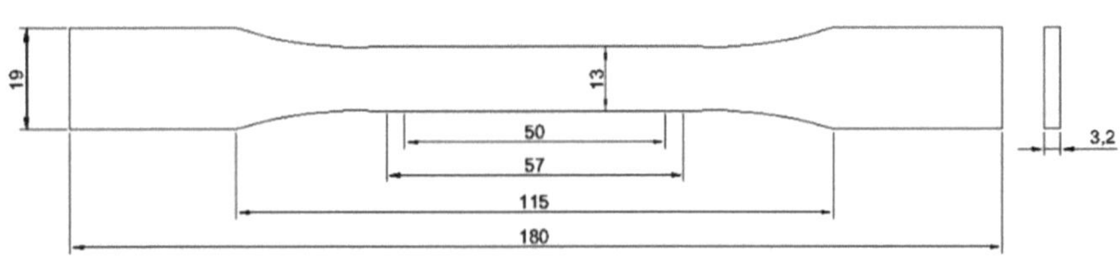

Fig. 5 Tensile specimen as per ASTM D638

Table 1 Build parameters and their variations

Fixed build parameters			Variable build parameters						
Parameters	Value	Unit	Parameters	Levels					Unit
				1	2	3	4	5	
Liquefier temperature	210	°C	Raster angle (RA)	0	30	45	60	90	°
Bed temperature	70	°C	Layer height (LH)	100	150	200	250	300	μm
Scan speed	50	mm/s	Raster width (RW)	400	500	600	700	–	μm
No of perimeters	1	–							
% Infill	100	%							
Infill pattern	Rectilinear	–							

Fig. 6 Experimental set up (OMEGA dual extruder)

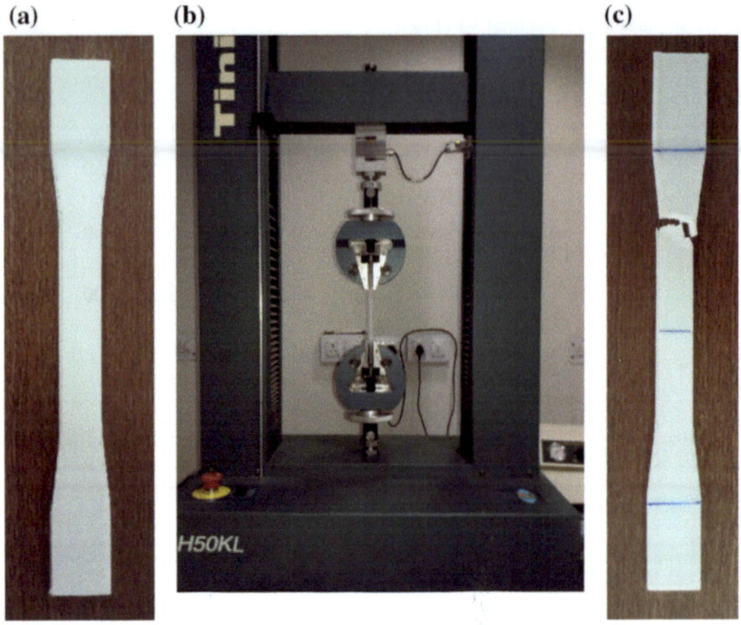

Fig. 7 Tensile test sample **a** before testing, **b** during testing and **c** after testing

according to ASTM D638. Crosshead motion will continue until the specimen breaks during the test. Figure 7 illustrate the tensile test sample before, during and afterward tensile testing.

3.4 ANFIS Model for Tensile Strength Prediction

The Neuro Fuzzy design tool of MATLAB Mathwork has been used to model ANFIS structure. This tool is provided for constructing and evaluating a fuzzy system using a GUI. User can use the ANFIS editor's GUI menu bar to load a new Sugeno system that loads the initialization of the FIS training, saves the trained FIS, and interprets the trained models. Figure 8 shows ANFIS structure used in the present study.

The training strategy for the ANFIS system is summarized in the flowchart of Fig. 9. This process begins with acquiring a training data set and verifying the dataset. The training data set is used to search the constant of the membership function. An error threshold between the current output and the desired output is determined.

In this study, the result was randomly divided into two parts. It is 80% for the training data (80 data points) of the learning process and 20% for the test data (20 data points).

Using the MATLAB software, the ANFIS model was developed from experimental data and the tensile strength was predicted. A fuzzy inference system of the

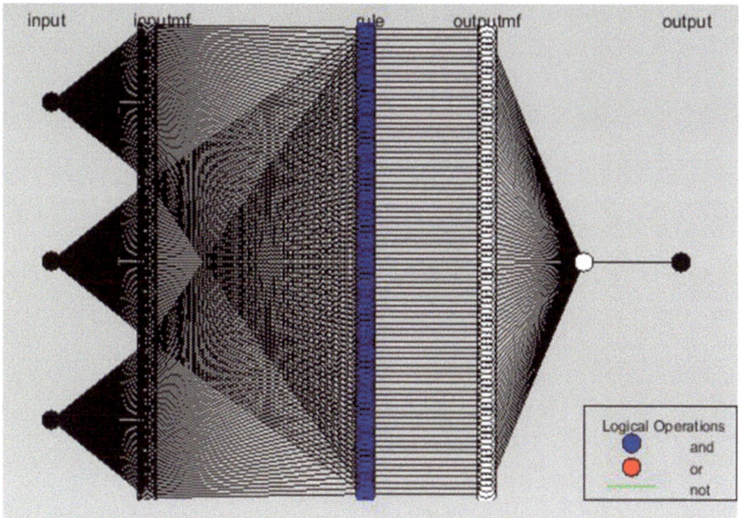

Fig. 8 Proposed ANFIS architecture

Sugeno type has been used for modeling the tensile strength. In order to predict the optimal selection of the Sugeno type FIS parameter, a hybrid learning algorithm was used. Due to their small error, 50 epochs were chosen.

The ANFIS training process begins with the determination of the membership functions. All learning data carried out by neural network, adjust the input parameters to the input-output relationship, and minimizes the error.

The parameters have been optimized after several tests. Finally, the model has been validated. The characteristics of the model are shown in Table 2.

4 Results and Discussion

In this study, all experiments were performed for each experimental design. It is desirable to maximize the tensile strength. The main effects graph is created using the average S/N ratio of each parameter at all levels. The maximum value of the average S/N ratio of the parameters is the best combination of parameters. The main effect is the direct influence of independent parameters on tensile strength. Figure 10 shows the effect of build variables on the tensile strength of FDM specimen. Figure 10 dissipates that tensile strength found to be decrease with increment in raster angle and layer height and increases when raster width is increased up to 600 μm. The analysis of S/N ratio revealed that the optimal tensile strength can be observed at raster angle 0° (Level 1), Layer height 100 μm (Level 1) and raster width 600 μm (Level 3).

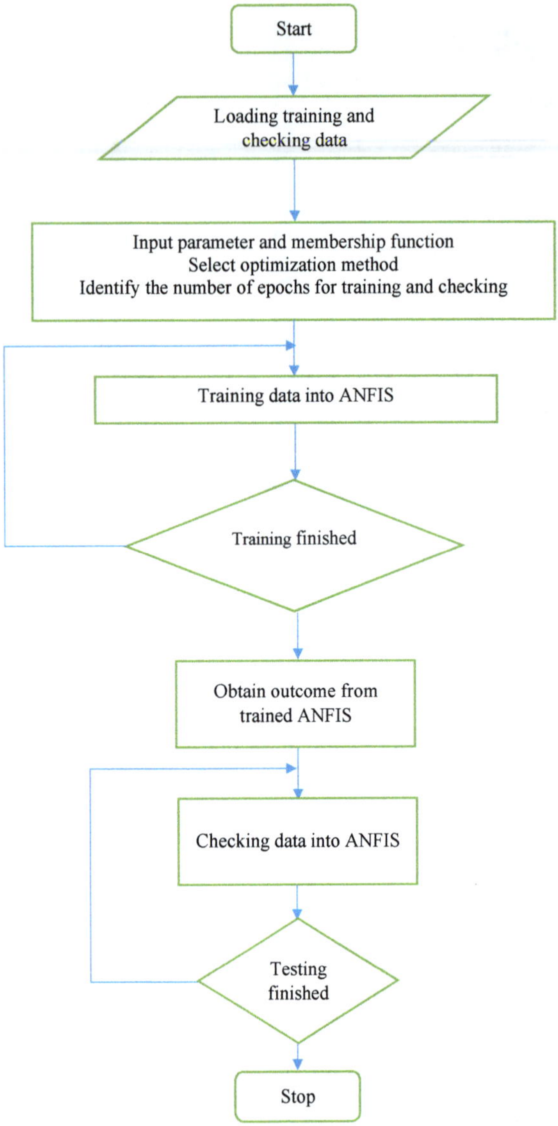

Fig. 9 ANFIS training system

Table 2 Characteristics of the developed ANFIS model

Parameter	Description/value
Structure of FIS	Sugeno
Initial FIS for training	genfis2 (subtractive clustering)
Range of influence	0.3
Squash factor	0.2
Accept ratio	0.2
Reject ratio	0.15
Number of inputs	3
Number of output	1
Number of input membership function	80 80 80
Optimization method	Hybrid
Training epoch number	50

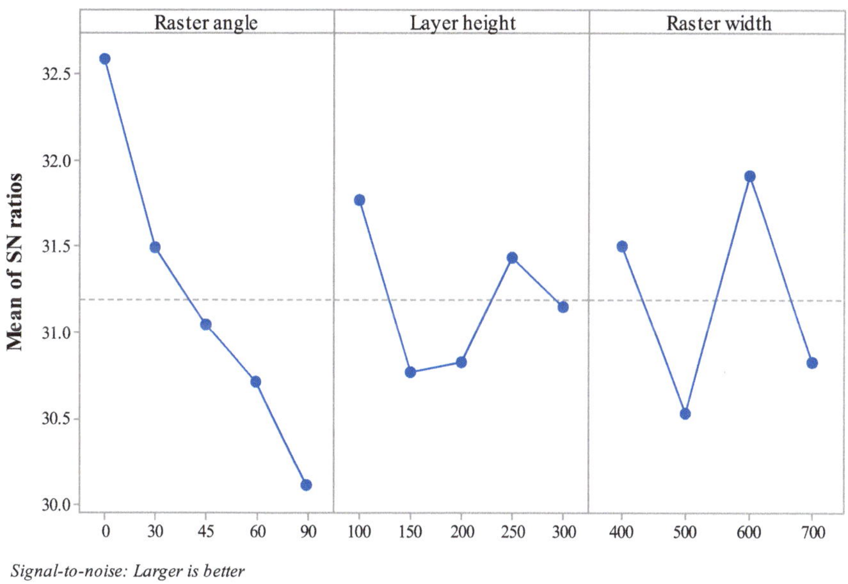

Fig. 10 Main effect plot of SN ratio for tensile strength

4.1 Evaluation of ANFIS Model Performance

Experiments were executed and the size of the test is determined by the classification of accuracy. The data is divided into two distinct sets of training data sets and checking data sets. The checking data was used to form ANFIS using the training dataset and to check the accuracy and efficiency of the training content of the ANFIS model trained for adaptation.

The performance of the ANFIS model was assessed using the coefficient of correlation (R^2), mean square error (MSE), root mean square error (RMSE), mean absolute error (MAE) and mean absolute percentage error (MAPE) [19, 20]. The R^2, MSE, RMSE, MAE and MAPE can be determined as follows.

$$MAPE = \frac{1}{n}\sum_{t=1}^{n}\left|\frac{E_t - P_t}{E_t}\right| \tag{11}$$

$$RMSE = \sqrt{\frac{\sum_{t=1}^{n}(\mathrm{E}_t - P_t)^2}{N}} \tag{12}$$

$$MSE = \frac{\sum_{t=1}^{n}(\mathrm{E}_t - P_t)^2}{N} \tag{13}$$

$$MAD = \frac{1}{n}\sum_{t=1}^{n}|P_t - E_t| \tag{14}$$

$$R^2 = \frac{\sum_{t=1}^{n}(\mathrm{E}_t - \overline{\mathrm{E}}_t)(\mathrm{P}_t - \overline{P}_t)}{\sqrt{\sum_{t=1}^{n}(\mathrm{E}_t - \overline{\mathrm{E}}_t)^2\sum_{t=1}^{n}(\mathrm{P}_t - \overline{P}_t)^2}} \tag{15}$$

where, E_t is experimental value, P_t is predicted value, n is the total number of data, $\overline{\mathrm{E}}_t$ is the mean of the experimental value and $\overline{P}_t$ is the mean of the predicted values.

The smaller the MAPE value, the better the predictability of the output. Table 3 represent the results of error measurements performed to evaluate prediction capability using training data and model test data. It can be seen that the error value is lower. Therefore, reasonable prediction accuracy is expected.

Experimental strength is compared with those of the predicated strength for train data and test data (as shown in Figs. 11 and 12). A High relative of coefficient suggest the predictive model to estimate the tensile strength for FDM fabricated part.

Similarly, Fig. 13 shows the difference between experimental and predicted results for test data and it can be observed that predicted strength is near to the experimental strength in most of the cases.

The variation in tensile strength with respect to build parameters are shown in Fig. 14. Figure 14a illustrate the relationship between layer height, raster angle and tensile strength. It can be seen that tensile strength is found to be reduced as the raster angle was increasing. The smaller value of the raster angle (0°) has a highest tensile strength. Moreover, Fig. 14b shows the relationship between layer height and raster width. At the higher value of raster width, tensile strength is found to be decreased

Table 3 Error measures for predicting the results

	MSE	RMSE	MAPE	MAE	R^2
Training	1.04015E−05	0.003226	9.8728E−05	0.002353	1
Testing	6.7319	2.5946	0.0644	2.2537	0.7946

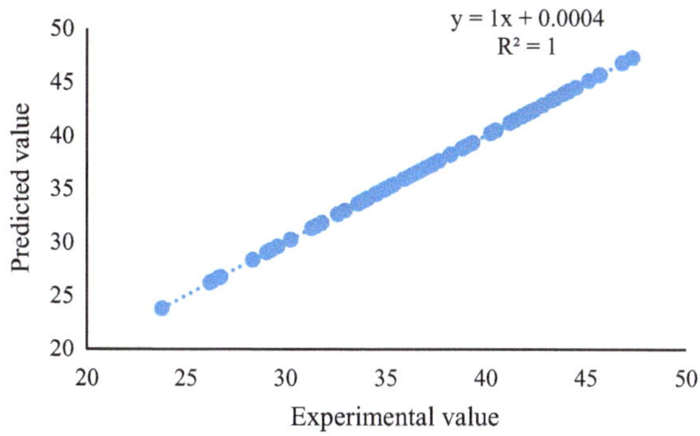

Fig. 11 Predicted versus experimental tensile strength during training using the ANFIS model

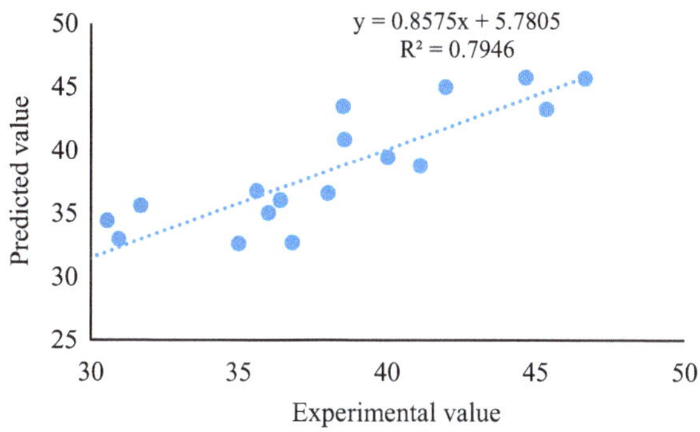

Fig. 12 Predicted versus experimental tensile strength during testing using the ANFIS model

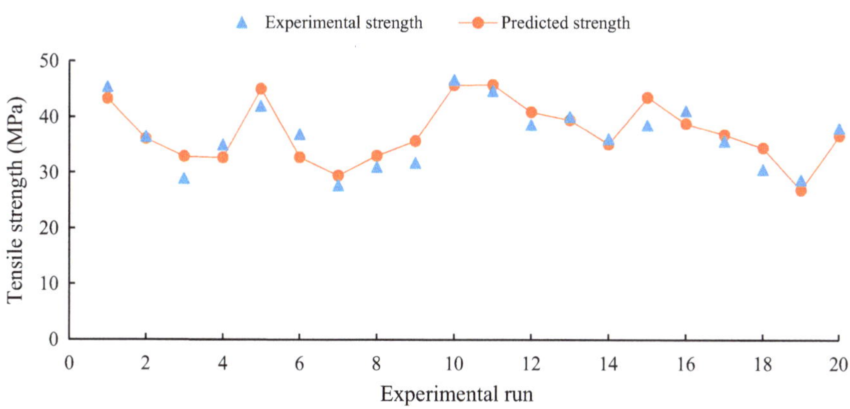

Fig. 13 The comparison of experimental and predicted tensile strength

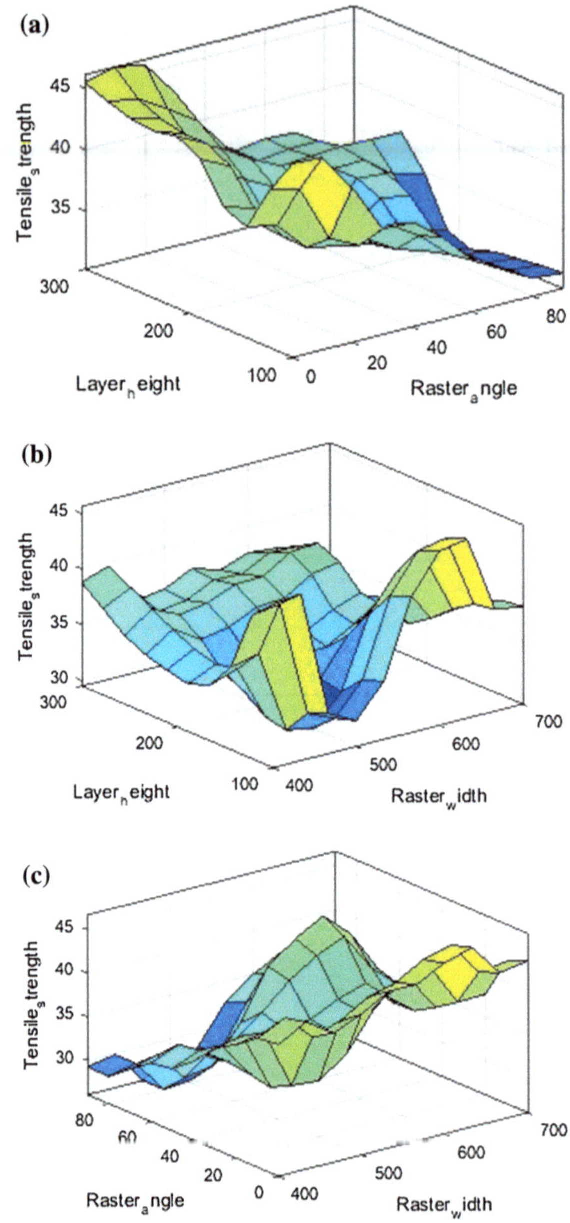

Fig. 14 Variation in tensile strength with respect to **a** layer height versus raster angle, **b** layer height versus raster width and **c** raster angle versus raster width

with increment in layer height. At the lower value of raster width, tensile strength is found to be decreased with increment in layer height and then its start increasing with further increment in layer height. Referring to Fig. 14c, it interprets that higher tensile strength is found to be achieved at 600 μm raster width and further its start decreasing.

5 Conclusion

The present work involves the development of an ANFIS prediction model for tensile strength in FDM printed part. The developed model allows for estimation of the tensile strength as a function of FDM build parameter which includes raster angle, layer height and raster width. The experimental response shows that build parameter deeply affects the tensile strength of FDM part. The developed model was verified for its predictability of the tensile strength using 20 samples of the testing data. ANFIS model constructed for predicting the tensile strength observed the result close to the experimental one. The model can be useful tool when involved in process planning to predict the tensile strength prior to manufacturing.

References

1. Gibson I, Rosen DW, Stucker B (2010) Additive manufacturing technologies, vol 238. Springer, New York
2. Gupta K (2018) Recent developments in additive manufacturing of gears: a review. In: Proceedings of 16th international conference on manufacturing research 2018, Skovde (Sweden), Advances in manufacturing technology XXXII, pp 131–136. IOS Press
3. Rajpurohit SR, Dave HK (2018) Effect of process parameters on tensile strength of FDM printed PLA part. Rapid Prototyp J. https://doi.org/10.1108/RPJ-06-2017-0134
4. Papazetis G, Vosniakos GC (2018) Mapping of deposition-stable and defect-free material extrusion additive manufacturing from minimal experiments. Int J Adv Manuf Technol. https://doi.org/10.1007/s00170-018-2820-1
5. Alimardani M, Toyserkani E (2008) Prediction of laser solid freeform fabrication using neuro-fuzzy method. Appl Soft Comput 8(1):316–323
6. Garg A, Tai K, Lee CH, Savalani MM (2014) A hybrid M5′-genetic programming approach for ensuring greater trustworthiness of prediction ability in modelling of FDM process. J Intell Manuf 25(6):1349–1365
7. Mohamed OA, Masood SH, Bhowmik JL (2017) Influence of processing parameters on creep and recovery behavior of FDM manufactured part using definitive screening design and ANN. Rapid Prototyp J 23(6):998–1010
8. Panda B, Shankhwar K, Garg A, Savalani MM (2016) Evaluation of genetic programming-based models for simulating bead dimensions in wire and arc additive manufacturing. J Intell Manuf 1–12
9. Panda BN, Bahubalendruni MR, Biswal BB (2014) Comparative evaluation of optimization algorithms at training of genetic programming for tensile strength prediction of FDM processed part. Procedia Mater Sci 5:2250–2257

10. Peng A, Xiao X, Yue R (2014) Process parameter optimization for fused deposition modeling using response surface methodology combined with fuzzy inference system. Int J Adv Manuf Technol 73(1–4):87–100
11. Raju M, Gupta MK, Bhanot N, Sharma VS (2018) A hybrid PSO–BFO evolutionary algorithm for optimization of fused deposition modelling process parameters. J Intell Manuf 1–16
12. Sood AK, Ohdar RK, Mahapatra SS (2010) Parametric appraisal of fused deposition modelling process using the grey Taguchi method. Proc Inst Mech Eng B J Eng Manuf 224(1):135–145
13. Sood AK, Ohdar RK, Mahapatra SS (2010) A hybrid ANN-BFOA approach for optimization of FDM process parameters. In: International conference on swarm, evolutionary, and memetic computing, Dec 2010, pp 396–403. Springer, Berlin, Heidelberg
14. Boschetto A, Giordano V, Veniali F (2013) Surface roughness prediction in fused deposition modelling by neural networks. Int J Adv Manuf Technol 67(9–12):2727–2742
15. Rayegani F, Onwubolu GC (2014) Fused deposition modelling (FDM) process parameter prediction and optimization using group method for data handling (GMDH) and differential evolution (DE). Int J Adv Manuf Technol 73(1–4):509–519
16. Jang JS, Sun CT, Mizutani E (1997) Neuro-fuzzy and soft computing—a computational approach to learning and machine intelligence. Prentice-Hall, Upper Saddle River, NJ
17. Suparta W, Alhasa KM (2016) Modeling of tropospheric delays using ANFIS. Springer International Publishing
18. Tan Y, Shuai C, Jiao L, Shen L (2017) An adaptive neuro-fuzzy inference system (ANFIS) approach for measuring country sustainability performance. Environ Impact Assess Rev 65:29–40
19. Şahin M, Erol R (2017) A comparative study of neural networks and ANFIS for forecasting attendance rate of soccer games. Math Comput Appl 22(4):43
20. Mashaly AF, Alazba AA (2018) ANFIS modeling and sensitivity analysis for estimating solar still productivity using measured operational and meteorological parameters. Water Sci Technol Water Supply 18(4):1437–1448

Optimization of Abrasive Water Jet Machining for Green Composites Using Multi-variant Hybrid Techniques

G. C. Manjunath Patel, Jagadish, Rajana Suresh Kumar and N. V. Swamy Naidu

Abstract Traditional machining of polymer matrix composites (PMCs) possesses difficulties as they exhibit excellent specific strength and stiffness. Superior properties led PMCs parts were extensively used in structural, aviation, construction and automotive applications. The advanced machining process abrasive water jet machining (AWJM) has been explored to machine PMCs. The AWJM factors namely abrasive grain size, working pressure, standoff distance, nozzle speed, and abrasive mass flow rate affect the final outcome of surface quality (i.e. surface roughness, SR) and productivity (i.e. material removal rate 'MRR' and process time 'PT') are studied. Taguchi L_{27} orthogonal array of experimental design is employed for conducting practical experiments. Taguchi method limit to optimize multiple conflicting outputs (maximize: MRR, and minimize: PT and SR), simultaneously. In general, multiple outputs may have many solutions and are dependent on the tradeoff (relative importance or weights) assigned to each output. Traditional practices such as engineer judgement, expert suggestion and customer requirements may lead to local solutions (i.e. superior quality for one output, while compromising with the rest). Principal component analysis (PCA) method overcomes the said shortcomings of traditional practices and determines weight fractions for each output based on the experimental data. Multi-objective optimization on the basis of ratio analysis (MOORA), Grey relational analysis (GRA), Technique for order preference by similarity to ideal solution (TOPSIS) and Data Envelopment Analysis based Ranking (DEAR) are the four methods employed for the purpose of multi-objective optimization. MOORA, GRA and TOPSIS methodologies require assigning weight fractions for each output by the problem solver. Note that, solution accuracies vary with the weight fractions assigned to each output. The aggregate (composite values of all responses) values determined by PCA-MOORA, PCA-TOPSIS, PCA-GRA and DEAR method were

G. C. Manjunath Patel
Department of Mechanical Engineering, P.E.S. Institute of Technology and Management, Shivamogga 577204, Karnataka, India

Jagadish (✉) · R. S. Kumar · N. V. S. Naidu
Department of Mechanical Engineering, National Institute of Technology Raipur, Raipur 492010, Chhattisgarh, India
e-mail: jagadishbaridabad.s@gmail.com

K. Gupta and M. K. Gupta (eds.), *Optimization of Manufacturing Processes*, Springer Series in Advanced Manufacturing, https://doi.org/10.1007/978-3-030-19638-7_6

used for determining optimal factor levels and their contributions. DEAR method determined optimal levels resulted in better machining quality characteristics.

Keywords Abrasive water jet machining · Optimization · PCA · MOORA · TOPSIS

1 Introduction

An organic polymer matrix which is used to bind fibers that are continuous is usually termed as polymer matrix composites (PMCs) [1]. PMCs can be categories into reinforced plastics and advanced composites. Due to their high specific strength and stiffness PMCs finds their application in wide areas of structural engineering namely aerospace, construction and automotive industries. However, superior strength and stiffness of the PMCs makes them very difficult to be machined by the conventional machining processes. Hence, to machine such materials which exhibit high specific strength and stiffness, Non-traditional methods of manufacturing play a vital role. Among all the available non-traditional methods of machining, Abrasive water jet machining (AWJM) has become the fastest growing method of non-traditional machining process due to its versatility [2]. Low cutting temperatures, presence of no heat activated zone (HAZ) on the material being cut, minimal dust and low cutting forces are the advantages it offers over its counterparts. AJWM also compliments its use with other non-traditional manufacturing technologies such as laser, EDM, and plasma etc. In AWJM, material removal takes place by impact energy developed over the surface to be machined using highly pressurized water containing abrasive particles. It makes this process a flexible machining method through which a wide range of higher strength materials can be machined.

The applicability of AWJM for milling of fiber reinforced plastics (FRP) was first carried out by Hocheng et al. where the authors analyzed the factor effects on MRR and SR in single pass cutting [3]. Arola and Ramulu used the micro analysis to know the material properties significance over surface integrity and texture [4]. Hloch et al. experimentally studied the cutting quality check and how the process parameters are influencing the same [5]. Analysis of variance (ANOVA) has been used to evaluate cutting quality making it a function of process parameters. Zhu et al. noticed that ductile erosion mechanism with small erosion angle and low pressure AWJM resulted in precise surface finish [6]. Selvan et al. considered SR as an important quality parameter, wherein good surface finish are obtained with more hydraulic pressure (P) and high abrasive flow rate (AFR) [7]. Manu and Babu observed that AWJM in turning can produce required turned surface by traversing the AWJ axially and radially while the workpiece is rotating [8]. The difficult to machine materials using AWJ turning has been studied and was found viable by Kartal and Gokkaya [9]. Borkowski [10] developed a novel mathematical model for the 3D sculpturing using a high pressure abrasive water jet by proposing an experimental test bed for shaping the materials. Wang [11] experimentally compared the various non-traditional machining methods and found that AWJM is best suited method to machine the polymer matrix composites. Muller and Monaghan [12] compared various non-traditional processes and wherein they concluded that AWJ machined part do not undergo problems associated to thermal damage. Siddiqui and Shukla [13] used Hybrid approach by combining the desired features of Taguchi method (TM) and PCA to assess the performance of AWJM by considering multiple quality characteristics (MQC). Aluminum and ferrous alloys find their application in many industries. Many studies are reported that optimizing the influencing variables results in economical machining for any alloy under AWJM. Iqbal et al. [14] used factorial experimental design to analyse the variable effects on maximum cutting width, machined surface texture, and percent of striation free area of AISI 4340 and Aluminum 2219.

1.1 Modelling and Optimization of AWJM Process

AWJM process performance depends on several process variables such as hydraulic pressure, work material, nozzle distance, abrasive type, size and mass flow rate etc. The research on AWJM process has been focused mostly on the process modelling and the optimization of the process parameters. Optimization of the process parameters in the AWJM process is of prime importance due to non-linear nature of the dependence of nozzle wear, kerf geometry, dimensional deviation, surface roughness, and MRR on process parameters [1, 2]. These factors regulate the performance of AWJM on the machinability of the material. Many research works reported on optimization of process based on statistical design of experiments (DOE) namely TM and response surface methodology (RSM) [15]. However, very little attention paid to model and optimize AWJM process by utilizing advanced tools namely GRA and soft computing tools etc. Azmir et al. used the grey rational analysis to optimize the control factors such as standoff distance (SoD), P and AFR on the Kevlar composite laminate surface finish [16]. Khan and Haque conducted experimental study to check the factor effects on the AWJ machined SR of glass fibre reinforced epoxy composites [17]. A linear regression equation representing SR as a mathematical function of process variables are derived by utilizing Taguchi method. Zohoor and Nourian [18] applied the response surface methodology to know the control factor effect of nozzle wear on the SR and developed regression equations. In addition, many research efforts were reported with a major focus on optimizing different factors for SR by utilizing TM, RSM and modern optimization tools [19–27]. Wang [27] presented the kerf quality of composite sheets (metal matrix) under AWJ machining. Shanmugam et al. [28] used kerf-taper compensation technique to minimize the kerf taper in AWJ cutting of alumina ceramics and found that compensational angle plays a major role on kerf taper. Srinivasu et al. [29] investigated the kinematic factor effects on the kerf geometry subjected to multi-jet erosion machining. The experimentation results formed a good basis for controlled three dimensional AWJ machining of complex geometries. In an important study, ANN model predicted better kerf geometry and SR on transformation induced plasticity of steel sheet [30]. For the last two decades researchers busy investigating and optimizing the effect of AWJM process parameters on MRR and nozzle wear using modern optimization techniques [27–34].

From the review of literature it has been noted that modern optimization techniques have been used significantly to know the effect of various control factors of AWJM. However, it has been found that very few works on modelling of AWJM using multi criteria optimization techniques have been reported and hence it finds a scope for further research. Since, AWJM has multiple process parameters, multi-criteria technique may be best suited for its modeling. The mathematical complexity and tedious nature of the steps involved in the approaches call for new methods for process optimization. Various optimization techniques like GRA-PCA, TOPSIS, DEAR, and MOORA approaches are recently developed methods available for optimization. Taguchi (L_{27}) orthogonal array method is incorporated in the experiment by varying some of the independent but critical parameters like abrasive grain size, nozzle speed (NS), working pressure, SoD, and AFR. Since adoption of optimal process parameters has seen a saturated amount of research mostly of which are single response problem, whereas complexity lies in optimizing the conflicting multiple outputs. Based on the original concept of TOPSIS approach, Ren et al. introduced a novel modified optimizing technique M-TOPSIS [35]. The drawback often encounter with the original TOPSIS method is the rank reversals and evaluation failure. Due to simple evaluation process in TOPSIS method, it's being widely used for some complex unconventional machining processes. A similar optimization has been done in wire electrical discharge machining by Gadakh [36]. Three different cases including variety of parameters have been selected followed by evaluation of those parameters using TOPSIS approach is presented in the study. A similarity to the past results so obtained such that TOPSIS method is more suitable for optimizing many multi criteria decision making problems in the current manufacturing processes. As discussed earlier about the availability of variety of optimization techniques, Taguchi-DEAR method is very simplest and efficient approach. It is proven to be the extensively accurate method to determine the optimal process parameters in manufacturing process. Muthuramalingam et al. [37] have analyzed the abrasive flow orientation process parameters in abrasive water jet machining under Taguchi-DEAR approach. Taguchi based L_9 orthogonal method has been implemented in the experimental trails in which liquid water pressure, feed rate, AFR, and SoD are the input factors. MRR and SR performance characteristics enhancement has been studied by the researchers using Taguchi-DEAR approach of solving MCDM problems and optimal process parameters has been computed. To obtain distinguishable physical properties, metal matrix composites reinforced with particles combined so as to form a more complex material are predominantly increasing which also provides high strength to the material. Machining such a material is a tedious job and therefore unconventional machining is being opted. Similar study reported by authors [38] with a focus on optimizing AWJM factors while machining TiB_2 particles reinforced Al7075 composites. Taguchi-DEAR methodology has been implemented to evaluate the performance measures such as MRR, taper angle and SR from the input parameters of water jet pressure, stand-off distance and transverse speed. Investigation results that water jet pressure predominantly affects the performance characteristics. The optimal process parameters are computed. Another optimization technique based on material selection is MOORA method. For designing any structure, the designers have to select materials with ultimate characteristics required for that particular design or structure. Inappropriate choice of material results in structure or design failure. In this diverse engineering world, fabrication of products corresponds to complex design demand for most challenging task of choice of appropriate materials for variety of components. Authors [39] reported the MOORA method is an appropriate tool for selection of proper materials. Various mathematical tools and techniques are found to be suitable for solving the material selection problems, which lead to the affected results based on the weights assigned to the considered selection criteria. MOORA method is simple to understand and employs suitable normalization procedure. Reference point approach has also been tested for the considered problem by the researchers. Results have been observed that all the methods generate approximately similar rankings correspond to the material alternatives. More robust and simple method as compared to other optimization methods as discussed in the above literatures is MOORA based Taguchi method. In another important work, multi response problem is converted into single response problem by integrating MOORA method with Taguchi method [40]. It is being noticed that the time consumed in the calculations of the steps is reduced by applying the proposed method. The hybrid MOORA-Taguchi method can solve successfully many multi-response problems [41]. TOPSIS method for multi response optimization of friction stir welding process variables was also used [41]. Materials like aluminium alloys and its composites which are difficult to weld are joined by friction stir welding. Researchers have accentuated on friction stir welding of aluminium matrix composite reinforced with silicon carbide particle. Since FSW incorporates a non-consuming revolving tool which is plunged into the verges of the material to be joined and progressed along the weld line, tool revolving speed, tool transverse speed and tool pin profile type of process variables are optimized with multiple responses such as percentage elongation, tensile strength and hardness of the material. It also leads weld joint of superior quality. Multiple response characteristics can be improved through optimization technique like TOPSIS is being revealed by the researchers in this study. As the growth of industrialization, machining operations are stressed to work with multiple ranges of materials which a traditional response approach cannot optimize easily. To handle such operations multi response techniques have been developed and one such technique is GRA-PCA multi response approach. It is traditional multi response technique which transforms multi quality response into single response. Researchers used this method in their study of optimizing aluminium alloy employing Taguchi method [42]. Observations were done on the quality values i.e. output parameters of aluminium alloy by optimizing the input parameters of a CNC end milling machine. Taguchi L_{27} orthogonal array approach has been implemented to select optimal parameters of the machine. Since there can be numerous parameters some of which are uncertain or having incomplete information. Here GRA-PCA provides efficient solution to this uncertainty. The researchers have observed with the help of GRA-PCA approach that speed, depth of cut and feed rate influence surface roughness and material removal rate significantly.

Although a great deal of research efforts reported to optimize the multiple responses, still industries are looking for simple, flexible, ease of understanding and implementation tools that optimize the manufacturing process. For the conflicting objective functions there exists a multiple solution which is dependent on the assigned relative importance (weights or trade-off) to individual objective function. Conventional practice of determining weights based on engineer judgement, expert recommendation, and customer requirements may lead to erroneous results. Although the above practice offers better individual output performance, but failed to provide solutions that satisfy all outputs. Thereby, determining a single set of input conditions that satisfy the conflicting outputs (maximize: MRR, and minimize: SR and PT) is considered a tedious task for industry personal. PCA convert multiple correlated responses into independent quality indices (i.e. single objective function) while solving multiple objective functions. Weights for individual output functions are determined using PCA. MOORA, TOPSIS and GRA require assigning weight fractions while converting the multiple responses to single objective functions for solving optimization. Note that, If PCA determine the eigen value greater than 1 for more than one output (i.e. principal component), then there is no procedure defined yet to select the weights to ascertain a feasible solution [43]. However, DEAR method does not require estimation of weight fraction to solve multiple objective optimization problems. In DEAR method the combination of target (actual) outputs are mapped into a ratio (i.e. weighted sum of outputs corresponds to larger-the-better divided by sum of weighted outputs representing lower-the-better) such that the computed values give ratio ranks that could help to determine the set of optimal factor levels [43]. In the present work, attempts are made to illustrate the tools (PCA-GRA, PCA-MOORA, PCA-TOPSIS and DEAR) involving mathematical computation that could help not only to optimize the AWJM process, but also the proposed methodology can be used by any industry personnel to solve the similar complex real-world practical problems.

2 Materials and Methods

2.1 *Material Preparation*

Sundi wood dust (SWD) possessing density of 0.779 g/cm^3 and particle size of approximately 600 μm is used as a reinforcement material for preparation of work specimen. Cellulose, glucomannan, xylem, and linen are the major constituent materials present in the work sample. In addition, 6% of filler material present in the composite matrix (94%) which is composed of epoxy (grade LY 556) possessing a density of 1.26 g/cm^3 and hardener (HY 951). Note that, resin and hardener proportion are maintained equal to 10:8 by wt. The mixture of SWD and matrix are mechanically stirred followed by pouring to vacuum glass chamber and allowed to set under ambient environment for a curing period of 24 h. The composite samples are prepared to the dimension of 180 mm × 140 mm × 6 mm (refer Fig. 1a, b). The prepared samples are subjected to perform machining.

2.2 Experimental Procedure

AWJM equipment (make: KMT Waterjet Systems) used for performing experiments is shown in Fig. 2. Five independent factors namely AGS, SoD, WP, AMFR and NS operating under three levels (Table 1) and Taguchi (L_{27}) is used to design and perform the experiments. During the experimentation, orifice diameter of 0.20 mm, nozzle

Fig. 1 **a** Sundi wood dust, **b** Sundi wood dust based polymer specimen

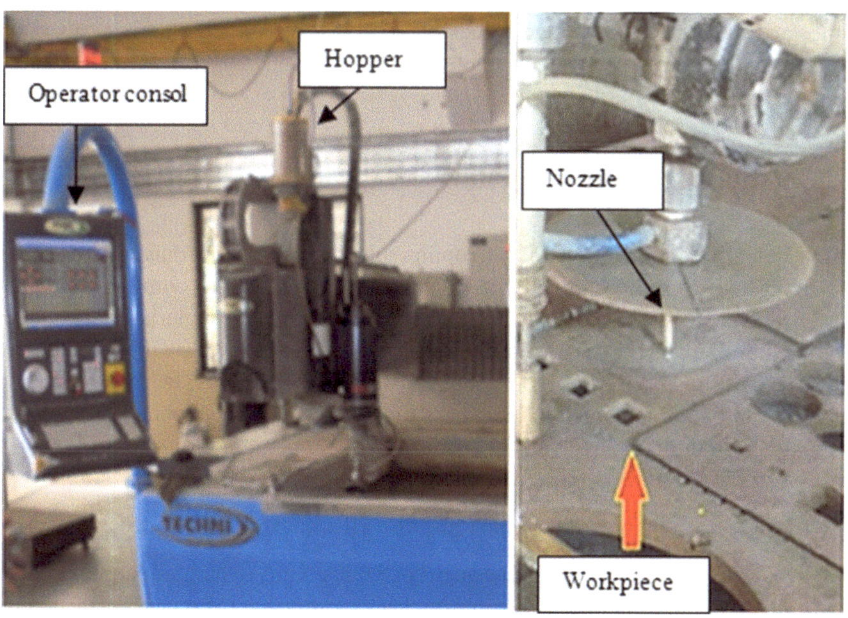

Fig. 2 **a** AWJM experimental setup, **b** AWJM nozzle head setup

Table 1 Input parameters and their levels for Taguchi design

Input parameters	Symbol	Units	Level 1	Level 2	Level 3
Abrasive grain size	AGS	mesh	60	80	100
Stand-off distance	SoD	mm	1.5	2.5	3.5
Working pressure	WP	MPa	150	200	250
Abrasive mass flow rate	AMFR	g/s	2	4	6
Nozzle speed	NS	mm/min	120	170	220
Constant parameters					
Orifice diameter	0.20 mm		Impact angle	90°	
Nozzle diameter	1.00 mm		Work piece thickness	5 mm	

diameter as 1 mm, and impact angle as 90° are used. For all experimental trials, the voltage and current are maintained equal to 300 V and 20 A. During experimentation the square holes of dimension (15 mm × 15 mm) are machined on the prepared green composite by using AWJM machine tool. Each experiment has been repeated three times and the measured average values of MRR, SR and process time are used for analysis and optimization (refer Table 2).

3 Methodology and Modelling

Multi-objective optimization refers to optimizing the process or product performance involving two or more outputs with or without the conflicting outputs simultaneously. The present work aims at simultaneously maximizing material removal rate while minimizing the process time and surface roughness of the AWJM process. The proposed offline optimization tools (PCA-GRA, PCA-MOORA, PCA-TOPSIS and DEAR) can be implemented in industries by any novice user to obtain the resulted benefits. Various steps involved for successful implementation of said tools with a case of AWJM process is presented in Fig. 3.

Step 1: Selection of input-output that improve efficiency of AWJM process

MRR, SR and PT are the important quality characteristics which affect the productivity, quality and economics while cutting PMCs using AWJM process. The quality characteristics are influenced directly by process parameters which affects the efficiency of AWJM process. For experimentation, analysis and optimization the most

Table 2 Experimental input-output data of AWJM process

Exp. no.	Input parameters					Output parameters			S/N ratio		
	AGS (mesh)	SoD (mm)	WP (MPa)	AMFR (g/s)	NS (mm/min)	MRR (mm^3/min)	PT (s)	SR (μm)	MRR	PT	SR
E1	60	1.5	150	2	120	110.763	0.497	0.151	40.89	6.07	16.42
E2	60	1.5	150	2	170	59.129	0.821	0.186	35.44	1.71	14.61
E3	60	1.5	150	2	220	71.762	0.701	0.198	37.12	3.09	14.07
E4	60	2.5	200	4	120	51.783	1.054	0.204	34.28	−0.46	13.81
E5	60	2.5	200	4	170	59.519	0.895	0.225	35.49	0.96	12.96
E6	60	2.5	200	4	220	61.949	0.626	0.187	35.84	4.07	14.56
E7	60	3.5	250	6	120	51.083	1.052	0.287	34.17	−0.44	10.84
E8	60	3.5	250	6	170	52.329	1.128	0.339	34.38	−1.05	9.40
E9	60	3.5	250	6	220	62.342	0.876	0.267	35.90	1.15	11.47
E10	80	1.5	200	6	120	69.296	0.757	0.398	36.81	2.42	8.00
E11	80	1.5	200	6	170	91.458	0.535	0.295	39.22	5.43	10.60
E12	80	1.5	200	6	220	99.469	0.547	0.392	39.96	5.24	8.13
E13	80	2.5	250	2	120	65.192	0.529	0.132	36.28	5.53	17.59
E14	80	2.5	250	2	170	104.381	0.452	0.281	40.37	6.90	11.03
E15	80	2.5	250	2	220	98.198	0.591	0.213	39.84	4.57	13.43
E16	80	3.5	150	4	120	119.327	0.457	0.172	41.54	6.80	15.29
E17	80	3.5	150	4	170	131.139	0.393	0.166	42.36	8.11	15.60
E18	80	3.5	150	4	220	96.234	0.436	0.201	39.67	7.21	13.94
E19	100	1.5	250	4	120	205.379	0.129	0.294	46.25	17.79	10.63

(continued)

Table 2 (continued)

Exp. no.	Input parameters					Output parameters			S/N ratio		
	AGS (mesh)	SoD (mm)	WP (MPa)	AMFR (g/s)	NS (mm/min)	MRR (mm^3/min)	PT (s)	SR (μm)	MRR	PT	SR
E20	100	1.5	250	4	170	176.260	0.257	0.291	44.92	11.80	10.72
E21	100	1.5	250	4	220	257.348	0.273	0.433	48.21	11.28	7.27
E22	100	2.5	150	6	120	191.215	0.347	0.293	45.63	09.19	10.66
E23	100	2.5	150	6	170	299.210	0.195	0.386	49.52	14.20	8.27
E24	100	2.5	150	6	220	267.542	0.213	0.313	48.55	13.43	10.09
E25	100	3.5	200	2	120	203.530	0.153	0.152	46.17	16.31	16.36
E26	100	3.5	200	2	170	341.325	0.152	0.273	50.67	16.36	11.28
E27	100	3.5	200	2	220	276.231	0.273	0.135	48.83	11.28	17.39

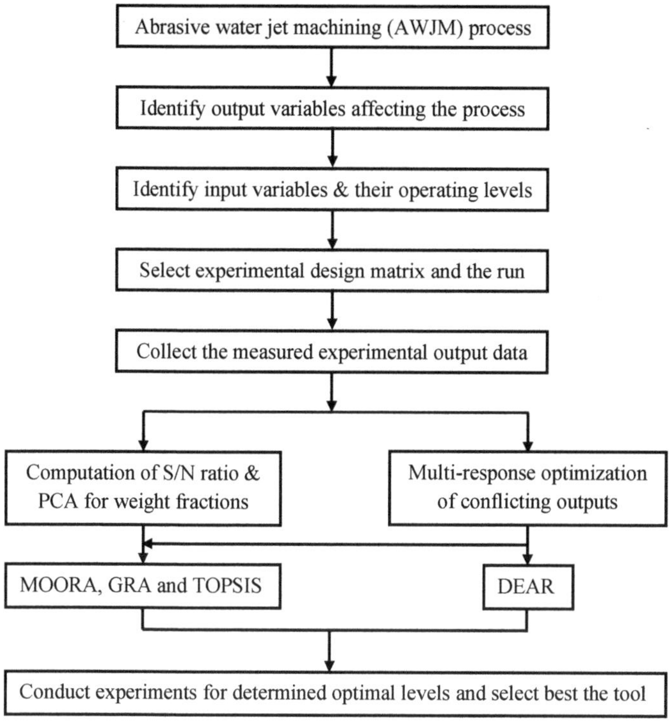

Fig. 3 Proposed steps for multi-response optimization in the present work

influencing parameters are selected based on consulting engineers and experts from industries, pilot experiment study results and available literatures [1, 3, 44]. Table 1 presents the influencing parameters and their operating levels used for experimentation.

Step 2: Selection of experimental plan and perform S/N ratio computation

Taguchi robust design was employed to conduct the experiments and perform statistical analysis. Taguchi method uses a special design of orthogonal arrays to investigate the entire factor space on performance characteristics with the set of minimum experimental trials. L_{27} orthogonal array experiments are employed for studying the influencing five factors operating with three different levels. Experiments are repeated three times for each run and the average values of measured performance characteristics of PT, MRR and SR are presented in Table 2. The experimental values of performance characteristics are converted to signal-to-noise (S/N) ratio. The present work involves two categories of quality characteristics when performing analysis with S/N ratio. Larger-the-better (LB) quality characteristics is employed for MRR, and smaller-the-better (SB) for SR and PT. Note that, S/N ratio depicted with higher values is treated as better quality characteristics irrespective of categories used. The signal-to-noise ratio computation corresponds to larger-the-better and smaller-the-better quality characteristics is done using Eqs. (1) and (2).

$$S/N_{SB} = -10\log\frac{1}{n}(y^2) \quad (1)$$

$$S/N_{LB} = -10\log\frac{1}{n}\left(\frac{1}{y^2}\right) \quad (2)$$

Terms, n corresponds to total number of experimental observations, and y represents the actual experimental data.

Step 3: Multi-response optimization of AWJM

The present work involves optimization of multiple responses which are conflicting in nature. Note that, multiple objective functions generate many solutions depending on relative importance (weight fractions) given to individual outputs, wherein each solution is different from one-another. This could occur due to the complex non-linear behavior of inputs towards outputs. Selecting the best solution among many potential solutions are treated as a tedious task for industry personnel. To limit the shortcomings of getting local solutions with traditional methods in deciding weight fractions for individual outputs, PCA was used.

3.1 Principal Component Analysis (PCA)

AWJM process requires optimization of multiple outputs. However, Taguchi robust design limit to optimize single output at once [45]. The goal of the present work is to locate the best set of control factors such that multiple responses are least sensitive to noise factors. PCA helps to determine the weight fractions for individual performance characteristics. The determined weight fractions correspond to each individual objective function was used to correlate the multiple outputs to single objective function while optimizing with MOORA, TOPSIS and GRA. Note that DEAR method does not require assigning weight fractions in their defined methodology.

The necessary steps essential to determine the weight fractions using principal component analysis are as follows:

1. Collection of output data

Let $X_i(j)$ corresponds to the experimental data. Where, i = (1, 2, 3, …, m) and j = (1, 2, 3, …, n). Terms, m and n represent the experimental run and quality characteristics.

2. Normalize the performance characteristics

Practical requirement suggested, smaller the better-quality characteristics for SR and PT (refer Eq. 3) and larger-the-better quality characteristics for MRR (refer Eq. 4). $X^*_{i,k}$ depicts the normalized data correspond to ith experiment and kth response.

For Smaller-the-better quality characteristics,

$$X^*_{i,k} = \frac{\min X_i(k)}{X_i(k)} \tag{3}$$

For Larger-the-better quality characteristics,

$$X^*_{i,k} = \frac{X_i(k)}{\max X_i(k)} \tag{4}$$

3. Computation of co-variance matrix.

V value corresponds to variance-covariance matrix that uses normalized data as discussed below,

$$V = \begin{bmatrix} V_{1,1} & V_{1,2} & \dots & V_{1,n} \\ V_{2,1} & V_{2,2} & \dots & V_{2,n} \\ \vdots & \vdots & \dots & \vdots \\ V_{m,1} & V_{m,2} & \dots & V_{m,n} \end{bmatrix} \tag{5}$$

where, $R_{i,j} = \frac{Cov\, X_i^*(j), X_i^*(k)}{\sigma X_i^*(j) * X_i^*(k)\theta} = \frac{Covariance\ of\ sequences\ X_i^*(j)\ and\ X_i^*(k)}{Standard\ deviation\ of\ sequences\ X_i^*(j)\ and\ X_i^*(k)}$

Step 4: Computation of eigen values and eigen vector of the covariance matrix

PCA was introduced to determine relative importance (weight fractions) for each performance characteristics (PT, SR, MRR). Minitab software platform is used for determining the weight fractions correspond to the performance characteristics shown in Table 4. The output data was used to estimate the correlation coefficient matrix which could help to estimate the Eigen values and Eigen vectors. The computed Eigen values and Eigen vectors are presented in Table 3.

The square value correspond to Eigen vector depicts the influence (i.e. significance) of each performance characteristic determined according to the principal component (refer Table 4). There are three outputs and hence three principal component values are determined. Note that, among all three principal components, the explained variance of the first principal component is as high as 62.3%. Important to note that the squares of first principal component eigen vectors are treated as weight fractions for the performance characteristics. The weight fractions associated to individual quality characteristics are found equal to 0.4942 for MRR, 0.4583 for PT and 0.0475 for SR, respectively.

Table 3 Eigen values and explained variation for principal components

Principal component	Eigen value	Explained variation (%)
First	1.8683	62.3*
Second	0.9815	32.7
Third	0.1502	05.0

*62.3 = 100 × 1.8683/(1.8683 + 0.9815 + 0.1502)

Table 4 Eigen vectors for principal components

Performance characteristics	Eigen vector			Weight fraction
	PC1	PC2	PC3	
Material removal rate, MRR	+0.703	0.035	0.710	0.4942*
Process time, PT	−0.677	−0.274	0.683	0.4583
Surface roughness, SR	+0.219	−0.961	−0.169	0.0475

0.4942* = 0.703 × 0.703

3.2 *Multi-objective Optimization on the Basis of Ratio Analysis (MOORA)*

In 2006, MOORA technique was developed by Brauers and Zavadskas. In general, MOORA methods work with the following three types namely, Ratio system, Reference point approach, and Full multiplicative form [40, 46]. The present work employed ratio system for the task optimization. The steps involved in Ratio System based MOORA are as discussed below [40]:

Step 1: Determination of decision matrix (D) wherein the characteristic values of alternatives at attributes η_{ij}. Terms, i = (1, 2, … m) and j = (1, 2, … r) are inputs represented in a matrix shown in Eq. (6).

$$D = \begin{bmatrix} \eta_{1,1} & \eta_{1,2} & \cdots & \eta_{1,r} \\ \eta_{2,1} & \eta_{2,2} & \cdots & \eta_{2,r} \\ \vdots & \vdots & \cdots & \vdots \\ \eta_{m,1} & \eta_{m,2} & \cdots & \eta_{m,r} \end{bmatrix} \tag{6}$$

Terms, m and r corresponds to total number of experimental observations or runs and number of responses, respectively.

Step 2: Computation of normalized and weighted normalized decision matrix

Equation (7) is used to calculate the normalized decision matrix. The mathematical formulation employed to calculate the weighted normalized decision matrix is presented in Eq. (8). In Eq. (8), w_j corresponds to the weight of the output or response j selected by the decision maker.

$$\eta_{ij}^{*} = \frac{\eta_{ij}}{\sqrt{\left(\sum_{i=1}^{m} (\eta_{ij})^2\right)}} \quad \mathrm{i} = 1, 2, \ldots \mathrm{m}; \mathrm{and\,j} = 1, 2, \ldots \mathrm{r} \tag{7}$$

$$Y_{ij} = \left[\eta_{ij} \times w_j\right]_{m \times r} \quad \mathrm{i} = 1, 2, \ldots \mathrm{m}; \mathrm{and\,j} = 1, 2, \ldots \mathrm{r} \tag{8}$$

η_{ij}^{*} is the normalized values of S/N ratio corresponding to i on response j.

Step 3: Computation of normalized and weighted normalized decision matrix

Y_i^* represents the ranking scores computation done by MOORA (refer Eq. 9). The computation of optimizing multiple conflicting responses use weighted normalized values correspond to maximize the better quality-characteristics and are subtracted with minimize the better quality-characteristics determine the MOORA index (Y_i^*).

$$Y_i^* = \underbrace{\sum_{j=1}^{l} Y_{ij}}_{\substack{maximize\,the\,better\\ quality\,charecteristics}} - \underbrace{\sum_{j=l+1}^{r} Y_{ij}}_{\substack{minimize\,the\,better\\ quality\,charecteristics}} \tag{9}$$

Terms in Eq. (9), here j = 1, 2, … l corresponds to number of responses to be maximized, and j = l + 1, l + 2, …, n represents the number of responses to be minimized. High value of Y_i^* is treated as better multiple quality characteristics.

Summary of Results of PCA-MOORA
PCA supply weights to MOORA that could optimize the multiple performance characteristics by determining the MOORA Index (Y_i^*). MOORA index Y_i^* values obtained from systematic procedure is used for further analysis and optimization.

3.3 Technique for Order Preference by Similarity to Ideal Solution (TOPSIS)

TOPSIS method estimates the solution by considering the shortest distance from the true solution (also called positive ideal solution), and farthest distance from negative true solution (ani-ideal solution) [47]. TOPSIS was developed in 1981 by Hwang and Yoon. In AWJM: True solution always aims at maximizing the MRR, and minimizing the SR and PT, whereas negative true solution maximizes the SR and PT and minimizes the MRR. TOPSIS work with the basic principle such that the best solution always lies, when it is closest to ideal solution and farthest to negative ideal solution. Important to note that, TOPSIS procedure does not give information about relative importance (weights) of those distances. Thereby, the relative importance is required for optimization of multiple outputs that are supplied with the help of PCA. The steps followed to optimize the multiple performance characteristics by utilizing PCA-TOPSIS are discussed below:

Step 1: Development of the decision matrix.

The decision matrix is composed of the S/N ratio of quality characteristics at responses (η_{ij}^*; where i = 1, 2, 3, … m and j = 1, 2, … r) are the inputs represented in decision matrix (D). In the present work, m corresponds to number of experimental observations = 27, and r represents number of responses = 3.

$$D = \begin{bmatrix} \eta_{1,1} & \eta_{1,2} & \cdots & \eta_{1,r} \\ \eta_{2,1} & \eta_{2,2} & \cdots & \eta_{2,r} \\ \vdots & \vdots & \cdots & \vdots \\ \eta_{m,1} & \eta_{m,2} & \cdots & \eta_{m,r} \end{bmatrix} \tag{10}$$

Step 2: Normalize the decision matrix.

The decision matrix is normalized according to Eq. (11). η_{ij}^* are the normalized values of S/N ratio corresponding to i on response j.

$$\eta_{ij}^* = \frac{\eta_{ij}}{\sqrt{\left(\sum_{i=1}^{m} (\eta_{ij})^2\right)}} \quad \mathrm{i} = 1, 2, \ldots \mathrm{m}; \mathrm{and\,j} = 1, 2, \ldots \mathrm{r} \tag{11}$$

Step 3: The computation of weighted normalized decision matrix is done according to Eq. (12).

$$V = [X_{ij}]_{m \times r} = [\eta_{ij} \times w_j]_{m \times r} \quad \mathrm{i} = 1, 2, \ldots \mathrm{m}; \mathrm{and\,j} = 1, 2, \ldots \mathrm{r} \tag{12}$$

w_j corresponds to weight fractions of jth response. $\sum_{j=1}^{r} w_j = w_1 + w_2 + \cdots w_r = 1$. Since, there are three outputs and weight fractions correspond to MRR, PT and SR is found equal to 0.4942, 0.4583 and 0.0475, respectively (refer Table 4).

Step 4: Calculate the positive ideal and anti-ideal (negative) solutions: A^* and A^- represents the ideal and negative ideal solution corresponds to maximum and minimum values of S/N ratio for all experimental trials (refer Eqs. 13–16).

$$A^* = (X_1^*, X_2^*, \ldots X_r^*) \tag{13}$$

$$X_j^* = \left[\left(\max_i X_{ij} \middle| j \in J\right) \quad \mathrm{i} = 1, 2, \ldots \mathrm{m}\right] \tag{14}$$

$$A^- = (X_1^-, X_2^-, \ldots X_r^-) \tag{15}$$

$$X_j^- = \left[\left(\min_i X_{ij} \middle| j \in J\right) \quad \mathrm{i} = 1, 2, \ldots \mathrm{m}\right] \tag{16}$$

Step 5: The calculation of d_i^* and d_i^- represents the ideal positive solution and ideal negative solution of distance of scenario i, respectively (refer Eqs. 17 and 18).

$$d_i^* = \sqrt{\sum_{j=1}^{r} \left(X_{ij} - X_j^*\right)^2} \quad \mathrm{i} = 1, 2, \ldots \mathrm{m}; \mathrm{j} = 1, 2, \ldots \mathrm{r} \tag{17}$$

$$d_i^- = \sqrt{\sum_{j=1}^{r}\left(X_{ij} - X_j^-\right)^2} \quad \mathrm{i} = 1, 2, \ldots \mathrm{m}; \mathrm{j} = 1, 2, \ldots \mathrm{r} \tag{18}$$

Step 6: Calculate the relative closeness or ranking score (C_i^*) of each alternative according to Eq. (19). C_i^* corresponds to larger the better quality-characteristics of alternative of A_i. Selection of the best alternative is decided based on the ranking score.

$$C_i^* = \frac{d_i^-}{\left(d_i^- + d_i^*\right)} \quad \mathrm{i} = 1, 2, \ldots \mathrm{m}; \mathrm{j} = 1, 2, \ldots \mathrm{r} \tag{19}$$

3.4 Grey Relational Analysis (GRA)

In 1982, Deng introduced the Grey Theory to handle poor, incomplete and uncertainty information. The grey color is neither black nor white [48]. In general, system is defined with color that represents the quantum of clear information (i.e. internal characteristics or mathematical formulations that details dynamics) about the system. If we know complete insight information about the system or process then it is called white system. Contrary, if the information is completely unknown then it is referred as the black system. Grey system refers to the information lies between the known and unknown information. The present work is based on the optimization of AWJM process, and maximizing the MRR and minimizing the SR and PT there. The steps followed for optimization using PCA-GRA are discussed below.

GRA is employed to calculate the relationship between reference (i.e. ideal) sequence $X_o^{(o)}(j)$ and comparable sequence $X_i^{(o)}(j)$, i = 1, 2, …, m; j = 1, 2, …, r, respectively.

Step 1: The S/N ratio values are computed for all responses including all experimental trials, $\left(\eta_{ij}\right)_{m \times r}$.
Step 2: Normalize the S/N ratio: S/N ratio values need to be normalized (using linear normalization) between the range of zero and one (unity). Note that the quality characteristics corresponds to larger-the-better and lower-the-better quality characteristics are computed using Eqs. (20) and (21).

$$Y_i(j) = \frac{\eta_i^o(j) - \min \eta_i^o(j)}{\max \eta_i^o(j) - \min \eta_i^o(j)} \tag{20}$$

$$Y_i(j) = \frac{\max \eta_i^o(j) - \eta_i^o(j)}{\max \eta_i^o(j) - \min \eta_i^o(j)} \tag{21}$$

Step 3: Calculate the deviation sequences as per Eq. (22). $\Delta_{oi}(j)$ is computed based on the absolute values of difference between reference sequence $x_o^*(j)$ and the comparable sequence of $x_i^*(j)$ after normalization.

$$\Delta_{oi}(k) = \left|Y_o^*(k) - Y_i^*(k)\right| \tag{22}$$

Step 4: Determine the grey relational coefficient (GRC). The purpose of GRC $\gamma(Y_o(j), Y_i(j))$ is to establish the relationship between the ideal and actual normalized S/N ratio for all responses.

$$\gamma(Y_o(j), Y_i(j)) = \frac{\Delta\min + \zeta\Delta\max}{\Delta_{oi}(j) + \Delta\max} \tag{23}$$

In general, the values correspond to $\Delta\max$, $\Delta\min$ and ζ is kept fixed to 1, 0 and 0.5 respectively.

Step 5: Calculate the overall performance by utilizing weighted grey relational grading (WGRG). The composite values of all responses associated with their respective weights determine the WGRG.

$$\gamma(Y_o, Y_i) = \sum_{j=1}^{r} w_1\left[\gamma(Y_o(j), Y_i(j))\right] + w_2\left[\gamma(Y_o(j), Y_i(j))\right] + \ldots w_r\left[\gamma(Y_o(j), Y_i(j))\right] \tag{24}$$

In the present work the required weights are supplied through PCA (refer Table 4). The weight fractions for MRR, PT and SR values are found equal to 0.4942, 0.4583 and 0.0475, respectively.

3.5 *Data Envelopment Analysis Based Ranking (DEAR)*

In 1978, Charnes et al. proposed the concept of data envelopment analysis (DEA). The DEA estimate the efficiency of a combination of decision-making units with utilization of multiple inputs to yield multiple outputs [49]. Note that, DEAR method used to solve for optimization of multiple responses does not require determination of weight fractions for individual quality characteristics. Here, set of actual outputs are correlated with simple mathematical formulation as a ratio such that the computed values estimate the ratio ranks. These ranks are further used for determining optimal factor levels and perform optimization. The sequential steps followed in DEAR for the estimation of multi-response performance index (MRPI) are:

Step 1: Calculate the weights (i.e. ratio of the performance measure at any trial to sum of all performance measures) for each output correspond to all experiments. The computation corresponds to determining the weight fraction of each output is done using the following Eqs. (25)–(27).

$$W_{mrr} = \frac{MRR}{\sum MRR} \tag{25}$$

$$W_{PT} = \frac{(1/PT)}{\sum (1/PT)} \tag{26}$$

$$W_{SR} = \frac{(1/SR)}{\sum (1/SR)} \tag{27}$$

Step 2: Transform the output data into weighted data after multiplying output with their corresponding weight fractions according to Eqs. (28)–(30).

$$M = \mathrm{W}_{mrr} \times \mathrm{MRR} \tag{28}$$

$$P = \mathrm{W}_{PT} \times \mathrm{PT} \tag{29}$$

$$S = \mathrm{W}_{SR} \times \mathrm{SR} \tag{30}$$

Step 3: Calculate the multi-performance ranking index by dividing the larger the better performance characteristics with smaller-the-better performance characteristics using Eq. (31).

$$MRPI = \frac{M}{P + S} \tag{31}$$

3.6 Determination of Optimal Factor Levels for All Outputs

The sets of optimal factor levels for abrasive water jet machining process are determined by applying multi-objective optimization tools (PCA-MOORA, PCA-TOPSIS, PCA-GRA, and DEAR). MOORA Index, TOPSIS ranking score, WGRG, and MRPI values represent the composite values correspond to all responses estimated by their methodology employed from PCA-MOORA (refer Table 5), PCA-TOPSIS (refer Table 6), PCA-GRA (refer Table 7), and DEAR (refer Table 8), respectively. Tables 9 and 10 present the consolidated MOORA Index, TOPSIS ranking score, WGRG, and MRPI of all factors operating at different levels. Example (say MOORA), the factors are calculated by adding all the MOORA index values operating under particular level of individual factors. It is worth mentioning that the choice of optimal level for a factor corresponds to the maximum level value of input factors on determining the performance characteristics. The optimal level of input factors of the AWJM process is determined by PCA-MOORA, PCA-TOPSIS, PCA-GRA, and DEAR method is presented in Tables 9 and 10. It is observed that, DEAR method determined optimal factor levels are different from those obtained for other methods studied. The rank for the factors were found to be different for different models and this might be due to the steps and procedure in determining the composite responses are found to be different. The higher difference (max. − min.) value corresponds to the individual factor resulted in highest importance (contribution or significance) on the performance measures. Abrasive grain size resulted in highest significance considering all the responses, as their corresponding difference value is more compared to other factors. Confirmation experiments are conducted for the determined optimal levels for a factor as obtained by PCA-MOORA, PCA-TOPSIS, PCA-GRA and DEAR, respectively. Note that the optimal factor levels determined for PCA-MOORA, PCA-TOPSIS and PCA-GRA are not among the combination of total twenty-seven experiments performed as per Taguchi method. This occurs due to the multi-facture nature of experimental design method (i.e. $3^5 = 243$ combinatorial set). This indicates that optimization methods (PCA-MOORA, PCA-GRA and PCA-TOPSIS) determined best factor levels is found to be one among the total set of 243 possible experimental conditions. DEAR method estimated optimal set of factor levels correspond to the 26th experimental trial (E_{26}) from the total 27 experiments conducted.

3.7 Confirmation Experiments

Confirmation experiments are conducted to verify the predictions of optimum techniques and to select the best optimization method for enhancing the multiple performance characteristics of AWJM process. Important to note that, DEAR method outperformed other methods (PCA-GRA, PCA-TOPSIS, and PCA-MOORA) in determining the optimal levels that resulted in desired high values of MRR, and low values of SR and PT. Note that, DEAR method produced 26.18% improvement in MRR, 17.83% for PT and 6.83% for SR, respectively (refer Table 11). Therefore, $A_3B_3C_2D_1E_2$ refers to the optimal set of factor levels recommended by DEAR method for AWJM process. The significance of individual factors was tested based on the obtained difference values of maximum and minimum levels. Abrasive grain size followed by nozzle speed, stand-off distance, working pressure and abrasive mass flow rate are the factors listed according to their importance in enhancing the multiple performance characteristics. The optimal factor levels of AWJM process is attributed to the following process mechanism. In AWJM, material removal phenomenon initiates with indentation on the work surface with the impact of abrasive particle. Indentation to possible material removal is dependent primarily on size of the abrasive particle striking the work surface. Tilly [50] explains decrease in the material removal was observed with smaller particle size as a result of less erosion on the machined surface area. Note that, small abrasive grit size particle poses lesser energy which is not sufficient enough to make larger damage (i.e. indentation) results

Table 5 MOORA based optimization results summary

Exp. no.	S/N ratio			Sum of squares			Normalization			Weighted normalization			MOORA index Y_i^*
	MRR	PT	SR	MRR	PT	SR	MRR	PT	SR	MRR	PT	SR	
E1	40.89	6.07	16.42	1672.0!	36.8!	269.6!	0.19*	0.13	0.25	0.094*	0.061*	0.012*	0.167*
E2	35.44	1.71	14.61	1256.0	2.9	213.5	0.16	0.04	0.22	0.081	0.017	0.011	0.109
E3	37.12	3.09	14.07	1377.9	9.5	198.0	0.17	0.07	0.21	0.085	0.031	0.010	0.126
E4	34.28	−0.46	13.81	1175.1	0.2	190.7	0.16	0.01	0.21	0.079	0.005	0.010	0.084
E5	35.49	0.96	12.96	1259.5	0.9	168.0	0.17	0.02	0.20	0.082	0.010	0.009	0.101
E6	35.84	4.07	14.56	1284.5	16.6	212.0	0.17	0.09	0.22	0.082	0.041	0.011	0.134
E7	34.17	−0.44	10.84	1167.6	0.2	117.5	0.16	0.01	0.16	0.079	0.004	0.008	0.082
E8	34.38	−1.05	9.40	1181.3	1.1	088.4	0.16	0.02	0.14	0.079	0.010	0.007	0.075
E9	35.90	1.15	11.47	1288.8	1.3	131.6	0.17	0.03	0.17	0.083	0.011	0.008	0.102
E10	36.81	2.42	8.00	1355.0	5.9	064.0	0.17	0.05	0.12	0.085	0.024	0.006	0.115
E11	39.22	5.43	10.60	1538.2	29.5	112.4	0.18	0.12	0.16	0.090	0.054	0.008	0.152
E12	39.96	5.24	8.13	1596.0	27.5	66.1	0.19	0.11	0.12	0.092	0.052	0.006	0.150
E13	36.28	5.53	17.59	1316.2	30.6	309.4	0.17	0.12	0.27	0.083	0.055	0.013	0.151
E14	40.37	6.90	11.03	1629.7	47.6	121.7	0.19	0.15	0.17	0.093	0.069	0.008	0.170
E15	39.84	4.57	13.43	1587.2	20.9	180.4	0.19	0.10	0.20	0.092	0.046	0.010	0.147
E16	41.54	6.80	15.29	1724.7	46.2	233.8	0.19	0.15	0.23	0.095	0.068	0.011	0.174
E17	42.36	8.11	15.60	1793.5	65.8	243.4	0.20	0.18	0.24	0.097	0.081	0.011	0.190
E18	39.67	7.21	13.94	1573.7	52.0	194.3	0.18	0.16	0.21	0.091	0.072	0.010	0.173
E19	46.25	17.79	10.63	2139.1	316.5	113.0	0.22	0.39	0.16	0.106	0.178	0.008	0.292
E20	44.92	11.80	10.72	2017.8	139.2	114.9	0.21	0.26	0.16	0.103	0.118	0.008	0.229

(continued)

Table 5 (continued)

Exp. no.	S/N ratio			Sum of squares			Normalization			Weighted normalization			MOORA index Y_i^*
	MRR	PT	SR	MRR	PT	SR	MRR	PT	SR	MRR	PT	SR	
E21	48.21	11.28	7.27	2324.2	127.2	052.9	0.22	0.25	0.11	0.111	0.113	0.005	0.229
E22	45.63	09.19	10.66	2082.1	84.5	113.6	0.21	0.20	0.16	0.105	0.092	0.008	0.204
E23	49.52	14.20	8.27	2452.2	201.6	068.4	0.23	0.31	0.13	0.114	0.142	0.006	0.262
E24	48.55	13.43	10.09	2357.1	180.4	101.8	0.23	0.29	0.15	0.112	0.134	0.007	0.253
E25	46.17	16.31	16.36	2131.7	266.0	267.6	0.21	0.36	0.25	0.106	0.163	0.012	0.281
E26	50.67	16.36	11.28	2566.4	267.6	127.2	0.24	0.36	0.17	0.116	0.163	0.008	0.288
E27	48.83	11.28	17.39	2384.4	127.2	302.4	0.23	0.25	0.26	0.112	0.113	0.013	0.237

Sum of squares are calculated from S/N ratio: $40.89 \times 40.89 = 1672^!$, $6.07 \times 6.07 = 36.8^!$, and $16.42 \times 16.42 = 269.6^!$
Normalization computation for MRR, SR and PT: [Ref. Eq. 7]
$0.19^* = (40.89)/\surd(1672 + 1256 + \ldots 2384.4) = 40.89/\surd(46231.9)$ and similarly for PT and SR
Computation of weighted normalization: $0.19 \times 0.4942 = 0.094^*$, $0.13 \times 0.4583 = 0.061^*$, $0.25 \times 0.0475 = 0.012^*$
Computation of MOORA Index Y_i^*: $0.094 + 0.061 + 0.012 = 0.167^*$

Table 6 Summary of TOPSIS based optimization results

Exp. no.	S/N ratio			Normalization			Weighted normalization			Solutions		TOPSIS C_i^*
	MRR	PT	SR	MRR	PT	SR	MRR	PT	SR	d_i^*	d_i^-	
E1	40.89	6.07	16.42	0.19	0.13	0.25	0.094	0.061	0.012	0.119[a]	0.073[b]	0.380[c]
E2	35.44	1.71	14.61	0.16	0.04	0.22	0.081	0.017	0.011	0.164	0.028	0.147
E3	37.12	3.09	14.07	0.17	0.07	0.21	0.085	0.031	0.010	0.150	0.042	0.219
E4	34.28	−0.46	13.81	0.16	0.01	0.21	0.079	−0.005	0.010	0.186	0.008	0.039
E5	35.49	0.96	12.96	0.17	0.02	0.20	0.082	0.010	0.009	0.172	0.021	0.108
E6	35.84	4.07	14.56	0.17	0.09	0.22	0.082	0.041	0.011	0.141	0.052	0.267
E7	34.17	−0.44	10.84	0.16	0.01	0.16	0.079	−0.004	0.008	0.186	0.007	0.034
E8	34.38	−1.05	9.40	0.16	0.02	0.14	0.079	−0.010	0.007	0.192	0.002	0.008
E9	35.90	1.15	11.47	0.17	0.03	0.17	0.083	0.011	0.008	0.170	0.023	0.117
E10	36.81	2.42	8.00	0.17	0.05	0.12	0.085	0.024	0.006	0.157	0.035	0.183
E11	39.22	5.43	10.60	0.18	0.12	0.16	0.090	0.054	0.008	0.126	0.066	0.342
E12	39.96	5.24	8.13	0.19	0.11	0.12	0.092	0.052	0.006	0.128	0.064	0.334
E13	36.28	5.53	17.59	0.17	0.12	0.27	0.083	0.055	0.013	0.127	0.066	0.343
E14	40.37	6.90	11.03	0.19	0.15	0.17	0.093	0.069	0.008	0.111	0.081	0.420
E15	39.84	4.57	13.43	0.19	0.10	0.20	0.092	0.046	0.010	0.134	0.058	0.301
E16	41.54	6.80	15.29	0.19	0.15	0.23	0.095	0.068	0.011	0.112	0.080	0.418
E17	42.36	8.11	15.60	0.20	0.18	0.24	0.097	0.081	0.011	0.099	0.094	0.487
E18	39.67	7.21	13.94	0.18	0.16	0.21	0.091	0.072	0.010	0.109	0.084	0.435
E19	46.25	17.79	10.63	0.22	0.39	0.16	0.106	0.178	0.008	0.011	0.190	0.944

(continued)

Table 6 (continued)

Exp. no.	S/N ratio			Normalization			Weighted normalization			Solutions		TOPSIS C_i^*
	MRR	PT	SR	MRR	PT	SR	MRR	PT	SR	d_i^*	d_i^-	
E20	44.92	11.80	10.72	0.21	0.26	0.16	0.103	0.118	0.008	0.061	0.131	0.680
E21	48.21	11.28	7.27	0.22	0.25-	0.11	0.111	0.113	0.005	0.066	0.127	0.660
E22	45.63	09.19	10.66	0.21	0.20	0.16	0.105	0.092	0.008	0.087	0.106	0.549
E23	49.52	14.20	8.27	0.23	0.31	0.13	0.114	0.142	0.006	0.037	0.156	0.810
E24	48.55	13.43	10.09	0.23	0.29	0.15	0.112	0.134	0.007	0.044	0.148	0.771
E25	46.17	16.31	16.36	0.21	0.36	0.25	0.106	0.163	0.012	0.018	0.176	0.907
E26	50.67	16.36	11.28	0.24	0.36	0.17	0.116	0.163	0.008	0.015	0.178	0.922
E27	48.83	11.28	17.39	0.23	0.25	0.26	0.112	0.113	0.013	0.065	0.128	0.662
Positive ideal solution, A^* (Eq. 13)							0.116	0.178	0.013			
Negative ideal solution, A^- (Eq. 15)							0.079	−0.01	0.005			

Procedure employed for computation of weighted normalization for MOORA remains same for TOPSIS

Positive ideal solution as per Eq. (17) $0.119 = \sqrt{[(0.094 - 0.116)^2 + (0.061 - 0.178)^2 + (0.012 - 0.013)^2]}$

Negative ideal solution as per Eq. (18) $0.073 = \sqrt{[(0.094 - 0.079)^2 + (0.061 - (-0.01))^2 + (0.012 - 0.05)^2]}$

Rank scores C_i^* as per Eq. (19): $0.073/(0.119 + 0.073) = 0.380$

Table 7 Results summary of GRA based optimization

Exp. no.	S/N ratio			Normalization			Grey relation coefficient (GRC)			WGRG $\gamma(Y_o, Y_i)$
	MRR	PT	SR	MRR	PT	SR	MRR	PT	SR	
E1	40.89	6.07	16.42	0.408[a]	0.378	0.887	0.458[b]	0.446	0.815	0.469
E2	35.44	1.71	14.61	0.077	0.146	0.711	0.351	0.369	0.634	0.373
E3	37.12	3.09	14.07	0.179	0.220	0.659	0.378	0.391	0.594	0.395
E4	34.28	−0.46	13.81	0.007	0.031	0.634	0.335	0.340	0.577	0.349
E5	35.49	0.96	12.96	0.080	0.107	0.551	0.352	0.359	0.527	0.364
E6	35.84	4.07	14.56	0.101	0.272	0.706	0.357	0.407	0.630	0.393
E7	34.17	−0.44	10.84	0.000	0.032	0.346	0.333	0.341	0.433	0.342
E8	34.38	−1.05	9.40	0.012	0.000	0.206	0.336	0.333	0.387	0.337
E9	35.90	1.15	11.47	0.105	0.117	0.407	0.358	0.361	0.457	0.365
E10	36.81	2.42	8.00	0.160	0.184	0.071	0.373	0.380	0.350	0.375
E11	39.22	5.43	10.60	0.306	0.344	0.323	0.419	0.433	0.425	0.426
E12	39.96	5.24	8.13	0.351	0.334	0.083	0.435	0.429	0.353	0.428
E13	36.28	5.53	17.59	0.128	0.349	1.000	0.364	0.435	1.000	0.427
E14	40.37	6.90	11.03	0.376	0.422	0.364	0.445	0.464	0.440	0.454
E15	39.84	4.57	13.43	0.344	0.298	0.597	0.432	0.416	0.554	0.431
E16	41.54	6.80	15.29	0.446	0.417	0.777	0.475	0.462	0.692	0.479
E17	42.36	8.11	15.60	0.496	0.486	0.807	0.498	0.493	0.722	0.507
E18	39.67	7.21	13.94	0.334	0.438	0.646	0.429	0.471	0.586	0.456
E19	46.25	17.79	10.63	0.733	1.000	0.326	0.652	1.000	0.426	0.801
E20	44.92	11.80	10.72	0.652	0.682	0.334	0.590	0.611	0.429	0.592
E21	48.21	11.28	7.27	0.851	0.654	0.000	0.771	0.591	0.333	0.668
E22	45.63	09.19	10.66	0.695	0.544	0.328	0.621	0.523	0.427	0.567
E23	49.52	14.20	8.27	0.931	0.809	0.097	0.879	0.724	0.356	0.783
E24	48.55	13.43	10.09	0.872	0.769	0.273	0.796	0.684	0.408	0.726
E25	46.17	16.31	16.36	0.728	0.921	0.881	0.647	0.864	0.808	0.755
E26	50.67	16.36	11.28	1.000	0.924	0.389	1.000	0.868	0.450	0.914
E27	48.83	11.28	17.39	0.889	0.654	0.981	0.818	0.591	0.963	0.722
Min.	34.17	−1.05	07.27							
Max.	50.66	17.79	17.59							

Normalization computation using Eq. [20–21]: [(40.89 − 34.17)]/[(50.66 − 34.17)] = 0.408
Computation of GRC using Eq. [23]: [0.0 + (0.5 × 1.0)]/[(1.0 − 0.408) + 0.5] = 0.458
Computation of WGRG using Eq. [24]: [(0.458 × 0.4942) + (0.446 × 0.4583) + (0.815 × 0.0475)] = 0.469

Table 8 Summary of DEAR based optimization results

Exp. no.	Experimental outputs			Weights for each output			Transform output data into weighted data			MRPI $M/(P+S)$
	MRR	PT	SR	W_{MRR}	W_{PT}	W_{SR}	M	P	S	
E1	110.763	0.497	0.151	0.030[a]	0.027[b]	0.056	3.340[c]	0.014	0.008	152.30[d]
E2	59.129	0.821	0.186	0.016	0.016	0.045	0.952	0.014	0.008	43.40
E3	71.762	0.701	0.198	0.020	0.019	0.042	1.402	0.014	0.008	63.93
E4	51.783	1.054	0.204	0.014	0.013	0.041	0.730	0.014	0.008	33.29
E5	59.519	0.895	0.225	0.016	0.015	0.037	0.964	0.014	0.008	43.98
E6	61.949	0.626	0.187	0.017	0.022	0.045	1.045	0.014	0.008	47.64
E7	51.083	1.052	0.287	0.014	0.013	0.029	0.710	0.014	0.008	32.39
E8	52.329	1.128	0.339	0.014	0.012	0.025	0.745	0.014	0.008	33.99
E9	62.342	0.876	0.267	0.017	0.015	0.031	1.058	0.014	0.008	48.25
E10	69.296	0.757	0.398	0.019	0.018	0.021	1.307	0.014	0.008	59.61
E11	91.458	0.535	0.295	0.025	0.025	0.028	2.277	0.014	0.008	103.84
E12	99.469	0.547	0.392	0.027	0.025	0.021	2.693	0.014	0.008	122.82
E13	65.192	0.529	0.132	0.018	0.026	0.064	1.157	0.014	0.008	52.76
E14	104.381	0.452	0.281	0.028	0.030	0.030	2.966	0.014	0.008	135.25
E15	098.198	0.591	0.213	0.027	0.023	0.039	2.625	0.014	0.008	119.71
E16	119.327	0.457	0.172	0.032	0.030	0.049	3.876	0.014	0.008	176.76
E17	131.139	0.393	0.166	0.036	0.034	0.051	4.682	0.014	0.008	213.49
E18	096.234	0.436	0.201	0.026	0.031	0.042	2.521	0.014	0.008	114.97
E19	205.379	0.129	0.294	0.056	0.105	0.029	11.483	0.014	0.008	523.63

(continued)

Table 8 (continued)

Exp. no.	Experimental outputs			Weights for each output			Transform output data into weighted data			MRPI $M/(P+S)$
	MRR	PT	SR	W_{MRR}	W_{PT}	W_{SR}	M	P	S	
E20	176.260	0.257	0.291	0.048	0.053	0.029	8.457	0.014	0.008	385.67
E21	257.348	0.273	0.433	0.070	0.050	0.019	18.029	0.014	0.008	822.15
E22	191.215	0.347	0.293	0.052	0.039	0.029	9.954	0.014	0.008	453.89
E23	299.210	0.195	0.386	0.081	0.069	0.022	24.372	0.014	0.008	1111.38
E24	267.542	0.213	0.313	0.073	0.064	0.027	19.486	0.014	0.008	888.57
E25	203.530	0.153	0.152	0.055	0.088	0.055	11.277	0.014	0.008	514.24
E26	341.325	0.152	0.273	0.093	0.089	0.031	31.715	0.014	0.008	1446.26
E27	276.231	0.273	0.135	0.075	0.050	0.062	20.772	0.014	0.008	947.23
Sum	3673.93									

Summation of 1/PT = 73.93 and 1/SR = 119.02
Weights for each output [Eqs. 25–27]: [110.763/3673.93] = 0.03015; [(1/0.497)/73.93] = 0.027
Transform output data into weighted data [Eqs. 28–30]: [0.03015 × 110.763] = 3.34
Computation of MRPI (refer Eq. 31): [3.340/(0.014 + 0.008)] = 152.30

Table 9 Response table for PCA-MOORA and PCA-TOPSIS

Variables	PCA-MOORA				PCA-TOPSIS			
	Level 1	Level 2	Level 3	Max. − Min.	Level 1	Level 2	Level 3	Max. − Min.
Abrasive grain size, A	0.980	1.422	**2.275**	1.295	1.320	3.264	**6.905**	5.585
Stand-off distance, B	1.568	1.505	**1.603**	0.098	3.890	3.608	**3.992**	0.383
Working pressure, C	**1.658**	1.541	1.477	0.182	**4.216**	3.766	3.508	0.709
Abrasive mass flow rate, D	**1.676**	1.605	1.395	0.281	**4.302**	4.038	3.150	1.152
Nozzle speed, E	1.550	**1.575**	1.552	0.025	3.798	**3.925**	3.767	0.159
Optimal levels / Rank	$A_3B_3C_1D_1E_2$		A > D > C > B > E		$A_3B_3C_1D_1E_2$		A > D > C > B > E	

The bold signifies the optimal levels for a factor

Table 10 Response table for PCA-GRA and PCA-DEAR

Variables	PCA-GRA				DEAR			
	Level 1	Level 2	Level 3	Max. − Min.	Level 1	Level 2	Level 3	Max. − Min.
Abrasive grain size, A	3.388	3.983	**6.527**	3.140	499.2	1099.2	**7093.2**	6593.8
Stand-off distance, B	4.527	4.494	**4.876**	0.381	2277.3	2886.5	**3527.6**	1250.2
Working pressure, C	**4.756**	4.726	4.416	0.339	3218.7	**3318.9**	2153.8	1165.1
Abrasive mass flow rate, D	**4.939**	4.609	4.350	0.590	**3475.1**	2361.6	2854.7	1113.5
Nozzle speed, E	4.565	**4.749**	4.584	0.185	1998.9	**3517.3**	3175.3	1518.4
Optimal levels / Rank	$A_3B_3C_1D_1E_2$		A > D > B > C > E		$A_3B_3C_2D_1E_2$		A > E > B > C > D	

The bold signifies the optimal levels for a factor

Table 11 Results of confirmation experiments tested for four optimization methods

Models	Optimal factor levels	Experimental performance characteristics
PCA-MOORA	$A_3B_3C_1D_1E_2$	MRR = 270.5 mm^3/min PT = 0.185 s SR = 0.293 μm
PCA-TOPSIS		
PCA-TOPSIS		
DEAR	$A_3B_3C_2D_1E_2$	MRR = 341.33 mm^3/min PT = 0.152 s SR = 0.273 μm

in less material and more cutting time. Increase in abrasive flow rate decreases the particle velocity and number of impacts as a result of increased interference between the particles [51]. Higher values of abrasive flowrate alter the impact angle of abrasive attack and reduce the local impact velocities which result in low material removal and increased process time. Lower the values of nozzle speed resulted in larger the depth of cut and better surface quality [52]. As the traverse speed increases, the depth of cut tends to decrease and favours for increased drag lines on the cut or machined surface resulted in rough machined surface. The jet diameter tends to expand with increased standoff distance [53], which favour the work piece exposed to larger machining area and the kinetic energy of abrasive particles strike the machining area at high impact with moderate work pressure resulted in better surface quality and productivity in machining.

4 Conclusions

In AWJM, machining parts to precise dimensional accuracy and surface finish is well established. Surface finish determines the functional performance characteristics of the machined parts, wherein its counterpart must not affect the productivity (i.e. MRR, and PT). Multi-objective optimization for the conflicting nature of outputs (i.e. minimize SR and PT, and maximize: MRR) are optimized for PCMs using AWJM process and the following conclusions can be drawn:

1. Taguchi robust design applied to conduct minimum practical experiments and collected the experimental input-output data. S/N ratio values are computed for the desired higher-the-better quality characteristics for MRR, and lower-the-better performance characteristics for SR and PT. Taguchi method collect output data and analyze the factors effects for each output separately, thus failed to optimize multiple outputs simultaneously.
2. Multiple outputs generate many solutions and are dependent on the nature of importance given to the response. Traditional practices (engineers or experts or customer advice) may yield the best output for one output, with the compromising solutions for the other. Statistical multi-variate analysis based principal compo-

nent analysis tool is used to determine the weight fractions based on the collected output data. PCA determined weight fractions for MRR, PT and SR values are found equal to 0.4942, 0.4583 and 0.0475, respectively. Note that, summation of all the weights correspond to the outputs must be maintained equal to one.
3. MOORA, TOPSIS and GRA require assigning weight fractions when performing multi-objective optimization. Thereby, PCA supply the determined weights to solve the said task. Note that, PCA-MOORA, PCA-TOPSIS, and PCA-GRA use different procedural steps to perform optimization. However, the recommended optimal levels ($A_3B_3C_1D_1E_2$) remain identical with slight change in ranking of factors.
4. DEAR method procedural steps itself estimate the weight fractions for each output at their respective experimental trials. Thus, the recommended optimal levels ($A_3B_3C_2D_1E_2$) and ranking of factors based on importance are different from those obtained from PCA-TOPSIS, PCA-MOORA, and PCA-GRA.
5. The confirmation experiments are conducted for the optimal levels suggested by all four methods. DEAR method outperformed other three models (PCA-GRA, PCA-TOPSIS and PCA-MOORA) in terms of yielding higher material removal rate with low process time and surface roughness. Abrasive grain size followed by nozzle speed, stand-off distance, working pressure and abrasive mass flow rate are the factors listed according to their importance in enhancing the multiple performance characteristics. Note that, DEAR method produced 26.18% improvement in MRR, 17.83% for PT and 6.83% for SR compared to other three (i.e. PCA-based models) models. This occurs due to each model possesses its own advantages and limitations with acceptable degree of errors in estimating values. Further, combining two such models do increase the computational complexity and time consuming. Therefore, DEAR method is a suitable tool which not only improves the product quality, but also provides solutions without much computation complexity and time. This could help any practice or novice engineer to apply tools for solving practical problems. Noteworthy, DEAR method can only optimize the conflicting nature of outputs is the only major limitation.

References

1. Jagadish KG, Rajkumaran M (2018) Evaluation of machining performance of pineapple filler based reinforced polymer composites using abrasive water jet machining process. In: Conference of the South African advanced materials initiative (CoSAAMI-2018). IOP Conf Ser Mater Sci Eng 430:012046
2. Gupta K, Gupta MK (2019) Developments in non-conventional machining for sustainable production—a state of art review. Proc Inst Mech Eng C J Mech Eng Sci. https://doi.org/10.1177/0954406218811982
3. Hocheng H, Tasi HY, Shiue JJ, Wang B (1997) Feasibility study of abrasive water jet milling of fiber rein forced plastics. J Manuf Sci Eng 119:133–142
4. Arola D, Ramulu M (1997) Material removal in abrasive waterjet machining of metals surface integrity and texture. Wear 210:50–58

5. Hloch S, Fabian S, Rimar M (2008) Design of experiments applied on abrasive waterjet factors sensitivity identification. Nonconv Technol Rev (2):49–57
6. Zhu HT, Huang CZ, Wang J, Li QL, Che CL (2009) Experimental study on abrasive waterjet polishing for hard–brittle materials. Int J Mach Tools Manuf 49:569–578
7. Selvan MCP, Raju NMS, Sachidananda HK (2012) Effects of process parameters on surface roughness in abrasive waterjet cutting of aluminum. Front Mech Eng 7:439–444
8. Manu R, Babu NR (2009) An erosion-based model for abrasive waterjet turning of ductile materials. Wear 266:1091–1097
9. Kartal F, Gokkaya H (2013) Turning with abrasive water jet machining—a review. Int J Eng Sci Technol 3:113–122
10. Borkowski PJ (2010) Application of abrasive-water jet technology for material sculpturing. Trans Can Soc Mech Eng 34:389–400
11. Wang J (1999) A machinability study of polymer matrix composites using abrasive waterjet cutting technology. J Mater Process Technol 94(1):30–35
12. Muller F, Monaghan J (2000) Non-conventional machining of particle reinforced metal matrix composite. Int J Mach Tools Manuf 40:1351–1366
13. Siddiqui TU, Shukla M (2008) Experimental investigation and hybrid multi-response robust parameter design in abrasive water jet machining of aircraft grade layered composites. IJAEA 1(5):39–48
14. Iqbal A, Dar NU, Hussain G (2011) Optimization of abrasive water jet cutting of ductile materials. J Wuhan Univ Technol Mater Sci Ed 26(1):88–92
15. Bhowmik S, Jagadish KG (2019) Modeling and optimization of advanced manufacturing processes. Springer, Switzerland. ISBN 978-3-030-00036-3
16. Azmir MA, Ahsan AK, Rahmah A, Noor MM, Aziz AA (2007) Optimization of abrasive waterjet machining process parameters using orthogonal array with grey relational analysis. In: Regional conference on engineering mathematics, mechanics, manufacturing & architecture (EM3ARC), pp 21–30
17. Khan AA, Haque MM (2007) Performance of different abrasive materials during abrasive water jet machining of glass. J Mater Process Technol 191:404–407
18. Zohoor M, Nourian SH (2012) Development of an algorithm for optimum control process to compensate the nozzle wear effect in cutting the hard and tough material using abrasive water jet cutting process. Int J Adv Manuf Technol 61(9–12):1019–1028
19. Chakravarthy PS, Babu NR (1999) A new approach for selection of optimal process parameters in abrasive water jet cutting. Mater Manuf Process 14(4):581–600
20. Chakravarthy PS, Babu NR (2000) A hybrid approach for selection of optimal process parameters in abrasive water jet cutting. J Eng Manuf 214:781–791
21. Wang J, Guo DM (2002) A predictive depth of penetration model for abrasive waterjet cutting of polymer matrix composites. J Mater Process Technol 121(2-3):390–394
22. Wang J (2007) Predictive depth of jet penetration models for abrasive waterjet cutting of alumina ceramics. Int J Mech Sci 49(3):306–316
23. Caydas U et al (2008) A study on surface roughness in abrasive waterjet machining process using artificial neural networks and regression analysis method. J Mater Process Technol 202:574–582
24. Jurkovic Z, Perinic M, Maricic S, Sekulic M, Mandic V (2012) Application of modeling and optimization methods in abrasive water jet machining. J Trends Dev Mach Assoc Technol 16(1):59–62
25. Mohamad A, Zain AM, Bazin NEN, Udin A (2015) A process prediction model based on Cuckoo algorithm for abrasive waterjet machining. J Intell Manuf 26(6):1247–1252
26. Parmar CM, Yogi PK, Parmar TD (2014) Optimization of abrasive water jet machine process parameter for AL-6351 using Taguchi method. Int J Adv Eng Res Dev (IJAERD) 1(5):1–8
27. Wang J (1999) A study of abrasive waterjet cutting of metallic coated sheet steels. Int J Mach Tools Manuf 39:855–870
28. Shanmugam DK, Wang J, Liu H (2008) Minimisation of kerf tapers in abrasive waterjet machining of alumina ceramics using a compensation technique. Int J Mach Tools Manuf 48:1527–1534

29. Srinivasu DS, Axinte DA, Shipway PH, Folkes J (2009) Influence of kinematic operating parameters on kerf geometry in abrasive water jet machining of silicon carbide ceramics. Int J Mach Tools Manuf 49:1077–1088
30. Vaxevanidis NM, Markopoulos A, Petropoulos G (2010) Artificial neural network modeling of surface quality characteristics in abrasive water jet machining of trip steel sheet. J Artif Intell Manuf Res 79–99
31. Nagdeve L, Chaturvedi V, Vimal J (2012) Implementation of Taguchi approach for optimization of abrasive water jet machining process parameters. Int J Instrum Control Autom (IJICA) 1:9–13
32. Aultrin KSJ, Anand MD, Jose PJ (2012) Modeling the cutting process and cutting performance in abrasive water jet machining using genetic-fuzzy approach. In: International conference on modeling optimization and computing (ICMOC-2012). Procedia Eng 38:4013–4020
33. Satyanarayana B, Srikar G (2014) Optimization of abrasive water jet machining process parameters using Taguchi grey relational analysis (TGRA). In: Proceedings of 13th IRF international conference, pp 135–140
34. Todkar M, Patkure J (2014) Fuzzy modelling and GA optimization for optimal selection of process parameters to maximize MRR in abrasive water jet machining. Int J Theor Appl Res Mech Eng 3(1):9–16
35. Ren L, Zhang Y, Wang Y, Sun Z (2007) Comparative analysis of a novel M-TOPSIS method and TOPSIS. Appl Math Res Express 2007, Article ID abm005, 10 pages
36. Gadakh VS (2012) Parametric optimization of wire electrical discharge machining using TOPSIS method. Adv Prod Eng Manag 7(3):157–164. ISSN 1854-6250
37. Muthuramalingam T, Vasanth S, Vinothkumar P, Geethapriyam T, Rabik MM (2018) Multi criteria decision making of abrasive flow oriented process parameters in abrasive water jet machining using Taguchi-DEAR methodology. Springer Science + Business Media B.V., Part of Springer Nature
38. Manoj M, Jinu GR, Muthuramalingam T (2018) Multi response optimization of AWJM process parameters on machining TiB_2 particles reinforced Al7075 composite using Taguchi-DEAR methodology. Springer Science + Business Media B.V., Part of Springer Nature
39. Karande P, Chakraborty S (2012) Application of multi-objective optimization on the basis of ratio analysis (MOORA) method for materials selection. Mater Des 37:317–324
40. Tansel Ic Y, Yildirim S (2013) MOORA-based Taguchi optimization for improving product or process quality. Int J Prod Res 51(11):3321–3341
41. Prabhu SR, Shettigar A, Herbert M, Rao S (2018) Multi response optimization of friction stir welding process variables using TOPSIS approach. IOP Conf Ser Mater Sci Eng 376:012134
42. Johny SKM, Sai CRS, Rao VR, Singh BG (2016) Multi-response optimization of aluminium alloy using GRA-PCA by employing Taguchi method. Int Res J Eng Technol 03(01)
43. Liao HC, Chen YK (2002) Optimizing multi-response problem in the Taguchi method by DEA based ranking method. Int J Qual Reliab Manage 19(7):825–837
44. Bhowmik S, Ray A (2016) Prediction and optimization of process parameters of green composites in AWJM process using response surface methodology. Int J Adv Manuf Technol 87(5–8):1359–1370
45. Chang HH (2008) A data mining approach to dynamic multiple responses in Taguchi experimental design. Expert Syst Appl 35(3):1095–1103
46. Kundakcı N (2016) Combined multi-criteria decision making approach based on MACBETH and MULTI-MOORA methods. Alphanumeric J 4(1):17–26
47. Opricovic S, Tzeng GH (2004) Compromise solution by MCDM methods: a comparative analysis of VIKOR and TOPSIS. Eur J Oper Res 156(2):445–455
48. Kayacan E, Ulutas B, Kaynak O (2010) Grey system theory-based models in time series prediction. Expert Syst Appl 37(2):1784–1789
49. Charnes A, Cooper WW, Rhodes E (1978) Measuring the efficiency of decision making units. Eur J Oper Res 2(6):429–444
50. Tilly GP (1973) A two stage mechanism of ductile erosion. Wear 23(1):87–96

51. Hashish M (1984) A modeling study of metal cutting with abrasive waterjets. J Eng Mater Technol 106(1):88–100
52. Kulekci MK (2002) Processes and apparatus developments in industrial waterjet applications. Int J Mach Tools Manuf 42(12):1297–1306
53. Chen FL, Siores E (2003) The effect of cutting jet variation on surface striation formation in abrasive water jet cutting. J Mater Process Technol 135(1):1–5

An Integrated Fuzzy-MOORA Method for the Selection of Optimal Parametric Combination in Turing of Commercially Pure Titanium

Akhtar Khan, Kalipada Maity and Durwesh Jhodkar

Abstract This chapter explores the application of a hybrid approach namely multi-objective optimization based on ratio analysis (MOORA) in fuzzy context to obtain the best parametric combination during machining of commercially pure titanium (CP-Ti) Grade 2 with uncoated carbide inserts in dry cutting environment. A series of experiment was performed by adopting Taguchi based L_{27} orthogonal array. Cutting speed, feed rate, and depth of cut were selected as three process variables whereas cutting force, surface roughness and flank wear were selected as three major quality attributes to be minimized. The minimization was exploited using fuzzy embedded MOORA method and hence an optimal parametric combination was attained. The results of the investigation clearly revealed that, the fuzzy coupled with MOORA method, was capable enough in acquiring the best parametric setting during turning operation under specified cutting conditions.

Keywords Fuzzy logic · MOORA · Flank wear · Optimization · Surface roughness · Titanium

1 Introduction

The present era is well-known for the creation and development of a large number of structural materials. Among these materials, titanium and its alloys are identified as more promising owing to their inherent properties. Titanium alloys are more

A. Khan (✉)
Department of Mechanical Engineering, Indian Institute of Information Technology, Design and Manufacturing, Kurnool 518002, Andhra Pradesh, India
e-mail: akhtarkhan00786@gmail.com

K. Maity
Department of Mechanical Engineering, National Institute of Technology Rourkela, Rourkela 769008, Odisha, India

D. Jhodkar
Department of Mechanical Engineering, G. H. Raisoni College of Engineering Nagpur, Nagpur 440016, Maharashtra, India

K. Gupta and M. K. Gupta (eds.), *Optimization of Manufacturing Processes*, Springer Series in Advanced Manufacturing, https://doi.org/10.1007/978-3-030-19638-7_7

attracting than that of other similar materials due to their unique characteristics such as superior corrosion resistance, highest strength-to-weight ratio, exceptional tissue-inertness and sustainability of these properties even at elevated temperatures [1–3]. Therefore, these alloys are most widely used in aerospace, chemical processing, marine, automobile and medical industries [4, 5]. Consequently, the aforementioned applications necessitate a substantial machining. Regrettably, titanium alloys are characterized as 'hard-to-cut' type materials owing to their poor thermal conductivity and high chemical affinity [6–8]. Poor thermal conductivity restricts heat dissipation from the primary cutting zone which in turn leads to excessive temperature gradient and hence to rapid tool wear at its pre-mature stages. Similarly, high chemical affinity of these alloys contributes in localizing the heat and hence a remarkable adhesion between tool and work materials which strongly curtails the tool life. The afore-discussed consequences may possibly lead to high production cost, high energy requirement in association with compromising dimensional accuracy [9]. These challenges can be addressed by selecting an appropriate cutting tool materials as well as a suitable combination of machining variables. However, a sufficiently enormous variety of cutting tool materials are now available, carbide inserts were recognized as the most suitable for machining titanium and its alloys [10, 11]. Therefore, commercially available uncoated carbide inserts were used for the machining of the selected work material during this investigation.

Selection of an appropriated combination of cutting variables is of paramount importance while machining "hard-to-cut" materials like titanium and its alloys. In machining of such alloys, an appreciable tool life, superior surface finish and relatively lower values of the cutting forces are acknowledged as the most noticeable manufacturing desires. To meet these requirements, adoption of optimization methods becomes essential to confirm high productivity without compromising the quality. In the past few decades, several experimental investigations have been reported exhibiting the application potential of different optimization techniques for optimizing turning parameters in order to achieve quality products. Lalwani et al. [12] studied the influence of various turning variables viz. cutting speed, feed rate and depth of cut, on two distinct cut qualities (i.e. cutting force and surface roughness) while turning MDN 250 steel with ceramic inserts. Aouici et al. [13], developed response surface methodology (RSM)-based quadratic models for the prediction of various turning responses such as surface roughness and cutting force, during turning of AISI H11 steel using CBN (Cubic boron nitride) inserts. Asiltürk and Neşeli [14] and Hashmi et al. [15], in their experimental investigations also developed RSM-based empirical models for the prediction of two different surface roughness characteristics viz. arithmetic mean roughness (R_a) and maximum peak-to-valley height (R_z). The results of both the investigations indicated that the suggested quadratic models were effective enough in predicting cut qualities and can be used to estimate the machining characteristics of other machining processes too. Similarly, Tebassi et al. [16] advised two different prediction models for estimating cutting force and surface roughness during machining of nickel based super alloy Inconel 718. They suggested RSM-based quadratic model and artificial neural network (ANN)-based model. In the current investigation, they compare the estimation efficiency of both the models and noticed that, ANN model was around 10.1% more precise in estimating cutting force (F_c) and 24.83% precise in estimating average roughness or arithmetic mean roughness (R_a), in comparison to the quadratic model counterpart. Furthermore, ANN model was also signified as an effective prediction tool for predicting cutting force, tool wear and surface roughness during turning operation [17]. Abburi and Dixit [18], recommended ANN coupled with fuzzy set theory to estimate the surface roughness during turning operation. Similarly, Basheer et al. [19], also used ANN model for predicting various output characteristics while precision machining of metal matrix composites (MMCs). The above mentioned studies were performed in dry cutting environment. In contrast, some researchers have reported the effectiveness of employing cooling media such minimum quantity lubrication (MQL) in order to attain improved machinability of different titanium and nickel-based super alloys [20–23]. These investigations highlighted the technical hitches and the benefits of MQL approach in a real time manufacturing system.

In addition to the afore discussed machining approaches, optimization techniques and prediction models, an extensive work has been found in which researchers have used various multi-criteria decision making (MCDM)-based approaches to solve turning problems consisting of multiple process variables and attributes. These methods include Analytical hierarchy process (AHP), Analytical network process (ANP), Fuzzy logic, multi-objective optimization based on ratio analysis (MOORA) method and Technique for Order of Preference by Similarity to Ideal Solution (TOPSIS) method etc. [24–28]. Apart from this, some hybrid and innovative techniques viz. Fuzzy-TOPSIS, Fuzzy-MOORA, and ANP-TOPSIS were also reported in order to confirm better solution of a specific problem [29]. The aforesaid approaches were found to be more precise in attaining the best alternative among a set of feasible alternatives. From the above literature survey it is clear that, an extensive work has been dedicated to the utilization of several MCDM-based approaches in solving a wide range of problems. However, turning parameters of CP-Ti grade 2 and their vagueness has not been studied and reported adequately so far. This might be contributed to the uncertain behavior of turning responses and vague information about the interaction between turning variables. Thus, the aforesaid situation leads an unclear solution. Therefore, selection of a suitable and effective methodology to solve MCDM-based problems is a great challenge to the researchers as well as the industries dealing with such situations. Keeping in mind, the vagueness and uncertainty of turning parameters, a fuzzy embedded MOORA method has been introduced in this study. The concepts of fuzzy set theory have been implemented to determine the best parametric combination while machining CP-Ti grade 2. The relationship between turning input and the selected output, were described with the help of fuzzy linguistic variables. In addition, a fuzzy control rule was developed for each of the selected attribute by adopting seven different linguistic grades. Thus, an optimal combination of process variables was attained and reported.

2 Multi-objective Optimization Based on Ratio Analysis (MOORA)

The MOORA method is a newly introduced approach having a substantial potential in dealing with a wide range of problems comprising of multiple as well as conflicting attributes. This method was developed and proposed by two European (Vilnius; Lithuania) researchers Brauers and Zavadskas in the year 2006. Basically, this method comprises of two distinct elements viz. the ratio system and the reference point approach. The first element is used to determine the overall performance of each alternative. This can be done by calculating the difference between the summations of the corresponding normalized values related to each criteria. On the other hand, the reference point approach helps in indicating the best or optimal combination of the alternatives.

The MOORA method can be understood deeply and clearly with the help of the following steps:

Step 1: Initially a decision matrix is constructed which represents all the selected responses and the corresponding set of input variables.

$$X = \begin{bmatrix} x_{11} & x_{12} & \dots & x_{1n} \\ x_{21} & x_{22} & \dots & x_{2n} \\ \vdots & \vdots & \vdots & \vdots \\ x_{m1} & x_{m2} & \dots & x_{mn} \end{bmatrix} \quad (1)$$

here, x_{ij} denotes the selected outcomes of the ith alternative on jth attribute, whereas m and n represents the number of alternatives (a set of input variables) and number of attributes (machining response) respectively.

Step 2: Normalization of the data sets observed in step number 1 and thus establishing a ratio system.

$$x_{ij}^{*} = \frac{x_{ij}}{\sqrt{\sum_{i=1}^{m} x_{ij}^{2}}} \quad (j = 1, 2, \dots, n) \quad (2)$$

here, x_{ij}^{*} denotes the normalized value of the ith alternative on jth attribute. This is a dimensionless quantity that lies between 0 and 1.

Step 3: In the next step, the overall assessment value is calculated by adding/subtracting the normalized values corresponding to each alternative. All the beneficial (higher-is-better) type performance characteristics are added whereas non-beneficial (lower-is-better are subtracted in order to obtain the overall assessment value.

$$y_i = \sum_{j=1}^{g} x_{ij}^{*} - \sum_{j=g+1}^{n} x_{ij}^{*} \quad (3)$$

here, g denotes the number of attributes related to beneficial criterion whereas $(n - g)$ is the number of attributes corresponding to non-beneficial criterion. y_i represents the overall assessment value of the ith alternative with respect to all alternatives.

Many a times, it was perceived that some of the attributes are of paramount importance when compared to the others. In such situations, weight criteria or factor can be multiplied with the same. After incorporation of weighting parameter the above equation can be written as below:

$$y_i = \sum_{j=1}^{g} w_j x_{ij}^* - \sum_{j=g+1}^{n} w_j x_{ij}^* \quad (4)$$

where, w_j is the weight of jth attribute.

Step 4: Assign ranking to overall assessment value y_i in descending order. The highest value of the y_i represents the best alternative, while the lowest value of y_i represents the worst.

3 An Introduction to Fuzzy Set Theory

A multi-criteria decision making (MCDM)—based problem has been identified as one of the difficult problems associated with real time manufacturing systems, due to the involvement of several uncertain situations. Therefore, to acquire an acceptable solution from this kind of problems has always been a challenging task to meet, for the researchers. In this situation, fuzzy set theory plays a key role in dealing with the ambiguity of the process and offers better results [30]. A fuzzy set theory, allows the decision maker to express their opinions in terms of specified linguistic variables. These linguistic variables can be converted into different fuzzy numbers with the help of fuzzy membership functions. In this way, MCDM-based problems can be solved easily and effectively. Consequently, fuzzy set theory has been identified as a significant as well as efficient tool for explaining human activities inclusive of vague and uncertain information.

Fuzzy inference system (FIS) is a well-recognized computing tool for handling linguistic knowledge and numerical data together. In general, FIS utilizes the concepts of fuzzy reasoning, fuzzy rules (If-then) and fuzzy set theory to deal with a wide range of problems viz. decision making, automatic control, robotics, classification of data and pattern recognition etc. This might be contributed to its effectiveness in mapping of any prescribed input to an output by using the aforesaid approaches. An FIS consists of four distinct elements such as fuzzifier, inference engine, knowledge base and defuzzifier. Initially, the crisp input is converted in terms of predefined linguistic variable by utilizing the membership function kept in the fuzzy knowledge base. This can be performed with the help of the first element i.e. fuzzifier and hence this process is termed as fuzzification. Secondly, the fuzzy input is converted to the

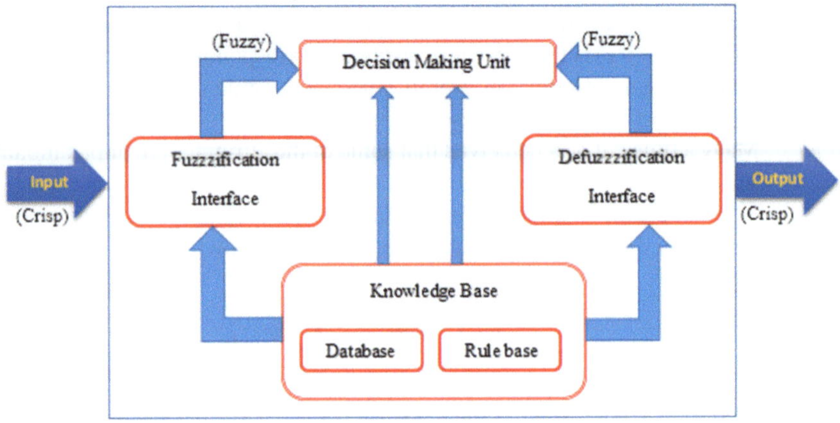

Fig. 1 The architecture of fuzzy—interface system

fuzzy output by adopting fuzzy rules (If-then) inside the inference engine. Finally, the last element i.e. defuzzified is engaged to convert these fuzzy output to a crisp value. The architecture of a fuzzy inference system is represented in Fig. 1.

Fuzzy number: A fuzzy number, is a subset of real numbers which denotes the development of the idea within a specified confidence interval [30]. For example, let A be the classical set of objects, whose elements are represented by X. The crisp value of a prescribed statement is characterized by means of a membership function and can be represented by a curve indicating the membership values lying in the range of 0 and 1.

$$\mu_A(X) = \begin{cases} 1, & if\ X \in A \\ 0, & Otherwise \end{cases} \tag{5}$$

Here, {0, 1} is known as the evaluation set and it is permissible to be represented in a real interval [0, 1] for the continuous mapping membership function. Moreover, assortment of a suitable membership function is of utmost significance in the fuzzification process. These membership functions are typically created by means of amply of elementary functions such as linear, quadratic and cubic polynomial curves, Gaussian distribution function, sigmoid curve etc. Conversely, the modest membership function can be created expending straight lines. In this category, the triangular membership function is recognized as the simplest one, which can be described with the help of a center-based triplet tactic. A triangular membership function can be constructed by keeping and equal and identical distance between the lowest and the highest points attached to the adjacent center. As a result of this, for each input value there will not be fuzzy sets greater than two. Similarly, the addition of their membership degrees always remains unity. Figure 2 explains the schematic representation of a triangular fuzzy membership function. For a clear understanding of fuzzy set theory and fuzzy numbers, some important definitions are enumerated below:

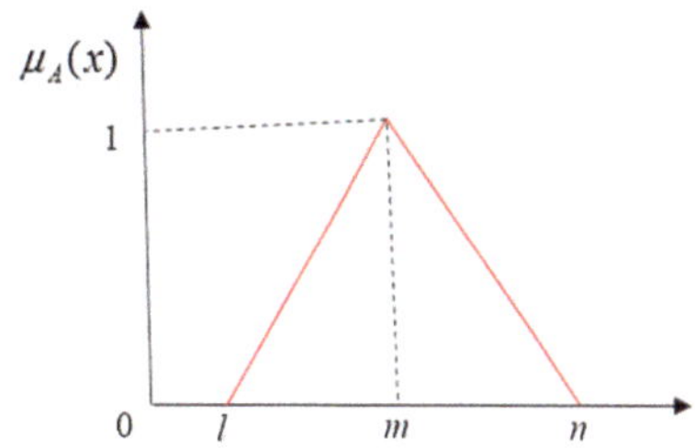

Fig. 2 A triangular fuzzy membership function

Definition 1: A fuzzy set $\tilde{A}$ in a universe of discourse X is described by a membership function $\mu_{\tilde{A}}(x)$ which is characterized as the grade of membership of x in $\tilde{A}$.

Definition 2: The triangular fuzzy numbers (TFNs) can be exemplified as $\tilde{A} = (a_1, a_2, a_3)$, and the membership function of the fuzzy number $\tilde{A}$ can be designated as below (Eq. 6):

$$\mu_{\tilde{A}}(x) = \begin{cases} 0 & x < 1, \\ \frac{x-a_1}{a_2-a_1} & a_1 \le x \le a_2, \\ \frac{a_3-x}{a_3-a_2} & a_2 \le x \le a_3, \\ 0 & x > a_3 \end{cases} \tag{6}$$

Definition 3: The fuzzy sum and fuzzy subtraction of two different TFNs are also triangular fuzzy numbers. But, the multiplication of two different TFNs is only an approximate TFN. For example, if there are two triangular fuzzy numbers $\tilde{A} = (a_1, a_2, a_3)$ and $\tilde{B} = (b_1, b_2, b_3)$, and a positive real number $r = (r, r, r)$, then the various algebraic operations between these two TFNs can be described as below:

$$\tilde{A}(+)\tilde{B} = (a_1 + a_2, b_1 + b_2, c_1 + c_2) \tag{7}$$

$$\tilde{A}(-)\tilde{B} = (a_1 - a_2, b_1 - b_2, c_1 - c_2) \tag{8}$$

$$\tilde{A}(\times)\tilde{B} = (a_1a_2, b_1b_2, c_1c_2) \tag{9}$$

$$\tilde{A}(/)\tilde{B} = (a_1/b_1, a_2/b_2, a_3/b_3) \tag{10}$$

$$\tilde{A}(\times)r = (a_1r, a_2r, a_3r) \tag{11}$$

Definition 4: The defuzzified value $m(\tilde{A})$ of a triangular fuzzy number $\tilde{A} = (a_1, a_2, a_3)$, can be calculated using Eq. (12):

$$m(\tilde{A}) = \frac{a_1 + a_2 + a_3}{3} \tag{12}$$

Definition 5: The distance between these two TFNs $\tilde{A} = (a_1, a_2, a_3)$ and $\tilde{B} = (b_1, b_2, b_3)$, can be calculated using Eq. (13):

$$d(\tilde{A}, \tilde{B}) = \sqrt{\frac{1}{3}(a_1 - b_1)^2 + (a_2 - b_2)^2 + (a_3 - b_3)^2} \tag{13}$$

Definition 6: The best non-fuzzy performance (BNP) value cab be computed by employing center of area (COA) method, as described in Eq. (14):

$$BNP_i = \frac{[(c - a) + (b - a)]}{3} + a, \forall_i \tag{14}$$

4 Fuzzy Embedded MOORA Method

The concept of fuzzy set theory in combination with MOORA method, was accomplished to estimate an optimal parametric combination in order to confirm better machinability of the selected work material. The hybridization of the two approaches attracted the attention of several researchers in the direction of the decision science community. Therefore, in the present work, an attempt has been made to exhibit the application potential of Fuzzy-MOORA method in solving an MCDM-based problem. The proposed hybrid approach offers a set of linguistic variables to express the opinions of decision makers. These variables were further utilized to construct fuzzy decision matrix and normalized fuzzy decision matrix. In the next step, weighted normalized matrix was acquired by adopting a suitable weightage for each of the selected response. Further, crisp values for weighted normalized fuzzy decision matrix was obtained, by calculating the best non-fuzzy performance value corresponding to each alternative. At the end, overall assessment values were computed and ranking was done by arranging them in descending order. The recommended hybrid approach consists of the following steps:

Step 1: Formation of fuzzy decision matrix using the adopted fuzzy triangular number illustrating all the alternatives (in rows) and attributes (in columns).

$$\tilde{X} = \begin{bmatrix} [x_{11}^l, x_{11}^m, x_{11}^n] & [x_{12}^l, x_{12}^m, x_{12}^n] & [x_{1n}^l, x_{1n}^m, x_{1n}^n] \\ \vdots & \vdots & \vdots \\ [x_{m1}^l, x_{m1}^m, x_{m1}^n] & x_{m2}^l, x_{m2}^m, x_{m2}^n & x_{mn}^l, x_{mn}^m, x_{mn}^n \end{bmatrix} \tag{15}$$

Step 2: Calculate the normalized fuzzy decision matrix using Eqs. (16–18).

$$x_{ij}^{l*} = \frac{x_{ij}^l}{\sqrt{\sum_{i=1}^m \left[\left(x_{ij}^l\right)^2 + \left(x_{ij}^m\right)^2 + \left(x_{ij}^n\right)^2\right]}} \tag{16}$$

$$x_{ij}^{m*} = \frac{x_{ij}^{m}}{\sqrt{\sum_{i=1}^{m}\left[\left(x_{ij}^{l}\right)^{2} + \left(x_{ij}^{m}\right)^{2} + \left(x_{ij}^{n}\right)^{2}\right]}} \quad (17)$$

$$x_{ij}^{n*} = \frac{x_{ij}^{n}}{\sqrt{\sum_{i=1}^{m}\left[\left(x_{ij}^{l}\right)^{2} + \left(x_{ij}^{m}\right)^{2} + \left(x_{ij}^{n}\right)^{2}\right]}} \quad (18)$$

Step 3: Estimate the weighted normalized fuzzy decision matrix using Eqs. (19–21).

$$V_{ij}^{m} = w_j x_{ij}^{m*} \quad (19)$$

$$V_{ij}^{l} = w_j x_{ij}^{l*} \quad (20)$$

$$V_{ij}^{n} = w_j x_{ij}^{n*} \quad (21)$$

here, w_j represents the weight criteria of each attribute.

Step 4: Convert the overall fuzzy assessment value ($\tilde{y}_i$) into a non-fuzzy value (crisp). The best non-fuzzy performance (BNP) can be calculated using Eq. (22).

$$BNP_i(y_i) = \frac{\left(y_i^n - y_i^l\right) + \left(y_i^m - y_i^l\right)}{3} + y_i^l \quad (22)$$

where $\tilde{y}_i = \left(y_i^l, y_i^m, y_i^n\right)$.

Step 5: Determine the overall fuzzy assessment value using Eq. (23).

$$\tilde{y}_i = \tilde{V}_{ij}^{+} - \tilde{V}_{ij}^{-} \quad (23)$$

here, $\tilde{V}_{ij}^{+}$ is the overall assessment value of beneficial criterion whereas $\tilde{V}_{ij}^{-}$ denotes the overall assessment value of non-beneficial criterion.

Step 6: Rank the above values by arranging them in descending order. The highest value exhibits the best alternative whereas the lowest value indicates the worst alternative.

5 Experimental Case Study

5.1 Work and Tool Materials

A cylindrical bar of commercially pure titanium (CP-Ti) was selected as the work material having diameter 50 mm and length 500 mm. The chemical composition

Table 1 Chemical composition of work material

Element	C	N	O	Fe	H	Ti
Wt. (%)	0.08–0.1	0.03	0.25	0.30	0.015	Balance

Table 2 Tool insert geometry

Parameter	
Insert shape	Square
Insert clearance angle	0°
Tolerance	±0.002
Cutting edge length	12 mm
Insert thickness	04 mm
Nose radius	0.8 mm
Holder style	PSBN
Shank height	20 mm
Shank width	20 mm
Tool length	125 mm

of the workpiece is listed in Table 1. A square shaped, ISO designated (SNMG 120408; Grade K313) cutting inserts were used for the machining of the selected work part. These inserts were rigidly mounted on a tool holder (ISO designation: PSBNR 2020K12). The geometry details of the cutting insert and tool holder are listed in Table 2.

5.2 Domain of the Investigation

The present investigation exploited Taguchi based orthogonal array design (L_{27}) to execute a series of experiment. These arrays were observed to be helpful in optimizing various quality characteristics and offering the best alternative amongst several alternatives. In the current investigation, an orthogonal array comprising of three factors and three levels is adopted as shown in Table 3. The allocation of the selected process variables was done according to the linear graph depicted in Fig. 3. Table 4, represents the experimental layout along with the measured outcomes.

5.3 Experimental Procedure

The selected round bar of the work material was turned on a heavy duty lathe (Model: NH-26; Manufacturer: HMT, India). A series of experiment was conducted as per

Table 3 Input variables with their levels

S. no.	Input variable	Unit	Levels		
			Level 1	Level 2	Level 3
1	Cutting speed (*v*)	m/min	30	60	90
2	Feed rate (*f*)	mm/rev	0.08	0.12	0.16
3	Depth of cut (*d*)	mm	0.2	0.4	0.6

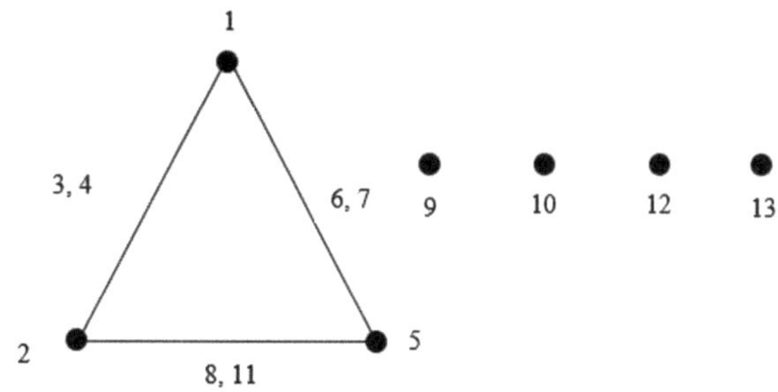

Fig. 3 Linear graph of proposed L_{27} orthogonal array

the list given in Table 4. Machining length was kept fixed as 250 mm and a new and sharp cutting edge was used for each experimental run. Figure 4 illustrates the experimental setup of the current investigation. Three distinct quality characteristics of turning operation viz. cutting force (F_c), surface roughness (R_a) and flank wear (VB) were examined and measured after completion of each trial. Cutting force was measured using a three dimensional (3D) piezoelectric dynamometer (Manufacturer: Kistler Instrument Corporation). The values of F_c were recorded at three different locations (roughly 80 mm apart) throughout the cutting length and the average value was noted. A roughness testing device (Model: Surtronic 3+, Manufacturer: Taylor Hobson) was used to measure the roughness parameter R_a of the machined surface. The measurements of R_a values were performed at six different locations (roughly 60° apart) around the circumference of the turned part. Similarly, wear on the flank surfaces of each cutting insert was examined and measured with the help of an optical microscope (Model: Axio Cam ER_c 5s, Manufacturer: Carl Zeiss). To confirm a better measurement accuracy, wear height at the flank surfaces of each cutting tool insert was recorded at three different locations and the average value was calculated for consideration.

Table 4 Outcomes of the experimentation

Run	Input variables			Responses		
	Speed	Feed	DOC	F_c (N)	R_a (μm)	VB (mm)
1	30	0.08	0.2	67.574	1.015	0.082
2	30	0.08	0.4	115.7	1.225	0.1
3	30	0.08	0.6	99.749	1.453	0.223
4	30	0.12	0.2	135.755	1.295	0.104
5	30	0.12	0.4	110.796	1.195	0.087
6	30	0.12	0.6	78.62	1.497	0.232
7	30	0.16	0.2	126.747	1.87	0.099
8	30	0.16	0.4	78.37	1.595	0.099
9	30	0.16	0.6	96.794	1.942	0.166
10	60	0.08	0.2	76.921	1.215	0.193
11	60	0.08	0.4	117.233	1.24	0.241
12	60	0.08	0.6	101.344	1.328	0.225
13	60	0.12	0.2	138.554	1.549	0.157
14	60	0.12	0.4	113	1.517	0.093
15	60	0.12	0.6	68.873	1.413	0.23
16	60	0.16	0.2	129.407	1.597	0.145
17	60	0.16	0.4	79.994	1.995	0.11
18	60	0.16	0.6	98.058	1.721	0.196
19	90	0.08	0.2	77.93	1.337	0.11
20	90	0.08	0.4	119.074	1.402	0.124
21	90	0.08	0.6	102.799	1.395	0.282
22	90	0.12	0.2	139.544	1.696	0.142
23	90	0.12	0.4	114.067	1.338	0.217
24	90	0.12	0.6	70.667	1.114	0.23
25	90	0.16	0.2	130.199	1.566	0.225
26	90	0.16	0.4	84.638	1.64	0.239
27	90	0.16	0.6	100.175	1.509	0.253

5.4 Estimation of Optimal Parametric Combination Using Fuzzy-MOORA Method

In the current investigation, fuzzy coupled with MOORA method was exploited to acquire the best parametric combination of input variables during machining of CP-Ti Grade 2 using uncoated carbide inserts in dry cutting environment. The main attention was given to minimize the cutting force and tool wear in combination with an appreciable surface finish. These performance characteristics are identified as

Fig. 4 Experimental setup

Table 5 Linguistic variables used for each criteria

Linguistic variable	Triangular fuzzy numbers (TFNs)
Very low (VL)	(0, 0, 0.1)
Low (L)	(0, 0.1, 0.3)
Medium low (ML)	(0.1, 0.3, 0.5)
Medium (M)	(0.3, 0.5, 0.7)
Medium high (MH)	(0.5, 0.7, 0.9)
High (H)	(0.7, 0.9, 1.0)
Very high (VH)	(0.9, 1.0, 1.0)

of paramount importance which significantly affect the production rate as well as the production cost, to a great extent. In this situation, acquiring and adopting the best parametric combination, become a challenging task. This also contributed to the vagueness of machining characteristics and the interaction effects among the selected process variables. Therefore, the proposed fuzzy set theory, uses linguistic terms such as very good, average, poor, very poor etc. for an effective assessment of the afore mentioned machining characteristics. Furthermore, the relative weights of each machining characteristic are also explained with the help of aforesaid fuzzy linguistic variables.

During this investigation, each alternative or experimental trail was primarily described in terms of specified linguistic variables as shown in Table 5. This was done to determine the relative weights of the selected output criterion viz. F_c, R_a and VB respectively, as listed in Table 6.

Secondly, valuation of all the available alternatives was accomplished based on the linguistic variables illustrated in Table 7. During this valuation, seven dissimilar fuzzy linguistic variables viz. very poor, poor, medium poor, fair, medium good, very good etc. were occupied. Table 8 represents the results of the assessment process.

Table 6 Relative weights of each criteria

Criteria	Decision maker	Fuzzy numbers
F_c	H	(0.7, 0.9, 1.0)
R_a	H	(0.7, 0.9, 1.0)
VB	VH	(0.9, 1.0, 1.0)

Table 7 Linguistic variables used for each alternative

Linguistic variable	Triangular fuzzy numbers (TFNs)
Very poor (VP)	(0, 0, 1)
Poor (P)	(0, 1, 3)
Medium poor (MP)	(1, 3, 5)
Fair (F)	(3, 5, 7)
Medium good (MG)	(5, 7, 9)
Good (G)	(7, 9, 10)
Very good (VG)	(9, 10, 10)

Further, formation of fuzzy decision matrix was done by converting the data sets attained after the aforesaid assessment process, into a suitable triangular fuzzy numbers. The results of the conversion process are depicted in Table 9.

Normalization of the data sets illustrated in fuzzy decision matrix (Table 9), was executed using Eqs. (16–18) and the outcomes are shown in Table 10. In the next step, the relevant weights of each machining criterion were multiplied with their corresponding values to attain weighted normalized fuzzy decision matrix as illustrated in Table 11.

The data sets listed in Table 11, were further converted into crisp values using Eq. (22) and depicted in Table 12. At the end, overall assessment values were calculated using Eq. (23) and listed in Table 13.

Finally, preference ranking was given to each alternative after arranging the overall assessment values in descending order, as exhibited in Table 13. By visualizing this table, it is clearly seen that, experiment number 1 is the best alternative offering minimum cutting force and tool wear along with appreciable surface quality. In contrast, experiment number 25, is signified as the worst alternative. Thus, an adequate machinability of the selected work material lies at lower range of machining variables. At lower, cutting speed, feed rate and depth of cut, machinability of the work part was observed to be better when compared to the higher ranges of machining variables counterpart. This might be contributed to the lower machining zone temperature at lower cutting speed, feed and depth of cut. Machining of titanium alloys at lower cutting speeds, does not raise the cutting zone temperature significantly whereas this temperature may be greater at high cutting speeds. High speed machining causes rapid growth in the cutting temperature which in turn introduces strain hardening and thermal softening phenomenon. This also results in a remarkable plastic deformation of the work part and curtails the machinability to a great extent. Therefore, lower range of the process variables are strongly recommended for

Table 8 Results of the assessment

Alternative	Responses			Fuzzy linguistic variables		
	F_c (N)	R_a (μm)	VB(mm)	F_c	R_a	VB
1	67.574	1.015	0.082	VG	VG	VG
2	115.7	1.225	0.1	MP	G	VG
3	99.749	1.453	0.223	F	F	MP
4	135.755	1.295	0.104	VP	MG	VG
5	110.796	1.195	0.087	MP	G	VG
6	78.62	1.497	0.232	G	F	P
7	126.747	1.87	0.099	VP	VP	VG
8	78.37	1.595	0.099	G	MP	VG
9	96.794	1.942	0.166	F	VP	MG
10	76.921	1.215	0.193	G	G	F
11	117.233	1.24	0.241	P	G	P
12	101.344	1.328	0.225	F	MG	MP
13	138.554	1.549	0.157	VP	F	MG
14	113	1.517	0.093	MP	F	VG
15	68.873	1.413	0.23	VG	MG	P
16	129.407	1.597	0.145	VP	MP	MG
17	79.994	1.995	0.11	G	VP	VG
18	98.058	1.721	0.196	F	P	F
19	77.93	1.337	0.11	G	MG	VG
20	119.074	1.402	0.124	P	MG	G
21	102.799	1.395	0.282	F	MG	VP
22	139.544	1.696	0.142	VP	MP	MG
23	114.067	1.338	0.217	MP	MG	MP
24	70.667	1.114	0.23	VG	VG	P
25	130.199	1.566	0.225	VP	F	MP
26	84.638	1.64	0.239	G	MP	P
27	100.175	1.509	0.253	F	F	P

machining titanium and its alloys, which is also witnessed during this investigation. However, this might be limited to the selected range of machining parameters and cutting conditions.

Table 9 Fuzzy decision matrix

Alternative	Responses		
	F_c	R_a	VB
1	9, 10, 10	9, 10, 10	9, 10, 10
2	1, 3, 5	7, 9, 10	9, 10, 10
3	3, 5, 7	3, 5, 7	1, 3, 5
4	0, 0, 1	5, 7, 9	9, 10, 10
5	1, 3, 5	7, 9, 10	9, 10, 10
6	7, 9, 10	3, 5, 7	0, 1, 3
7	0, 0, 1	0, 0, 1	9, 10, 10
8	7, 9, 10	1, 3, 5	9, 10, 10
9	3, 5, 7	0, 0, 1	5, 7, 9
10	7, 9, 10	7, 9, 10	3, 5, 7
11	0, 1, 3	7, 9, 10	0, 1, 3
12	3, 5, 7	5, 7, 9	1, 3, 5
13	0, 0, 1	3, 5, 7	5, 7, 9
14	1, 3, 5	3, 5, 7	9, 10, 10
15	9, 10, 10	5, 7, 9	0, 1, 3
16	0, 0, 1	1, 3, 5	5, 7, 9
17	7, 9, 10	0, 0, 1	9, 10, 10
18	3, 5, 7	0, 1, 3	3, 5, 7
19	7, 9, 10	5, 7, 9	9, 10, 10
20	0, 1, 3	5, 7, 9	7, 9, 10
21	3, 5, 7	5, 7, 9	0, 0, 1
22	0, 0, 1	1, 3, 5	5, 7, 9
23	1, 3, 5	5, 7, 9	1, 3, 5
24	9, 10, 10	9, 10, 10	0, 1, 3
25	0, 0, 1	3, 5, 7	1, 3, 5
26	7, 9, 10	1, 3, 5	0, 1, 3
27	3, 5, 7	3, 5, 7	0, 1, 3

6 Conclusions

In this chapter, an efficient and effective hybrid method has been projected to solve machining problems having multiple cut qualities under fuzzy environment. Selection of the best alternative in order to confirm better machinability of the selected work material, was done with the help of fuzzy embedded MOORA method. Thus in the present investigation a new MCDM approach, MOORA under fuzzy environment has been applied to deal with both quantitative and qualitative machining criteria. The following conclusions may be drawn after completion of current investigation:

Table 10 Normalized fuzzy decision matrix

Alternative	Responses		
	F_c	R_a	VB
1	0.9, 1.0, 1.0	0.9, 1.0, 1.0	0.9, 1.0, 1.0
2	0.1, 0.3, 0.5	0.7, 0.9, 1.0	0.9, 1.0, 1.0
3	0.3, 0.5, 0.7	0.3, 0.5, 0.7	0.1, 0.3, 0.5
4	0, 0, 0.1	0.5, 0.7, 0.9	0.9, 1.0, 1.0
5	0.1, 0.3, 0.5	0.7, 0.9, 1.0	0.9, 1.0, 1.0
6	0.7, 0.9, 1.0	0.3, 0.5, 0.7	0, 0.1, 0.3
7	0, 0, 0.1	0, 0, 0.1	0.9, 1.0, 1.0
8	0.7, 0.9, 1.0	0.1, 0.3, 0.5	0.9, 1.0, 1.0
9	0.3, 0.5, 0.7	0, 0, 0.1	0.5, 0.7, 0.9
10	0.7, 0.9, 1.0	0.7, 0.9, 1.0	0.3, 0.5, 0.7
11	0, 0.1, 0.3	0.7, 0.9, 1.0	0, 0.1, 0.3
12	0.3, 0.5, 0.7	0.5, 0.7, 0.9	0.1, 0.3, 0.5
13	0, 0, 0.1	0.3, 0.5, 0.7	0.5, 0.7, 0.9
14	0.1, 0.3, 0.5	0.3, 0.5, 0.7	0.9, 1.0, 1.0
15	0.9, 1.0, 1.0	0.5, 0.7, 0.9	0, 0.1, 0.3
16	0, 0, 0.1	0.1, 0.3, 0.5	0.5, 0.7, 0.9
17	0.7, 0.9, 1.0	0, 0, 0.1	0.9, 1.0, 1.0
18	0.3, 0.5, 0.7	0, 0.1, 0.3	0.3, 0.5, 0.7
19	0.7, 0.9, 1.0	0.5, 0.7, 0.9	0.9, 1.0, 1.0
20	0, 0.1, 0.3	0.5, 0.7, 0.9	0.7, 0.9, 1.0
21	0.3, 0.5, 0.7	0.5, 0.7, 0.9	0, 0, 0.1
22	0, 0, 0.1	0.1, 0.3, 0.5	0.5, 0.7, 0.9
23	0.1, 0.3, 0.5	0.5, 0.7, 0.9	0.1, 0.3, 0.5
24	0.9, 1.0, 1.0	0.9, 1.0, 1.0	0, 0.1, 0.3
25	0, 0, 0.1	0.3, 0.5, 0.7	0.1, 0.3, 0.5
26	0.7, 0.9, 1.0	0.1, 0.3, 0.5	0, 0.1, 0.3
27	0.3, 0.5, 0.7	0.3, 0.5, 0.7	0, 0.1, 0.3

- The best parametric combination to attain minimum cutting force, tool wear and surface roughness, was apparent at cutting speed 30 m/min, feed rate 0.08 mm/rev and depth of cut 0.2 mm, which was observed in experiment number 1.
- Lower surface roughness, cutting force and tool wear could be expected at moderate cutting speed, feed rate and depth of cut while machining CP-Ti grade 2 with uncoated carbide inserts under dry cutting environment.
- The proposed methodology was experienced systematic, easily understandable, and robust and can be implemented to solve similar types of problems associated in real time manufacturing systems.

Table 11 Weighted normalized fuzzy decision matrix

Alternative	Responses		
	F_c	R_a	VB
1	0.63, 0.9, 1.0	0.63, 0.9, 1.0	0.81, 1.0, 1.0
2	0.7, 0.27, 0.5	0.49, 0.81, 1.0	0.81, 1.0, 1.0
3	0.21, 0.45, 0.7	0.21, 0.45, 0.7	0.9, 0.3, 0.5
4	0, 0, 0.1	0.35, 0.63, 0.9	0.81, 1.0, 1.0
5	0.7, 0.27, 0.5	0.49, 0.81, 1.0	0.81, 1.0, 1.0
6	0.49, 0.81, 1.0	0.21, 0.45, 0.7	0, 0.1, 0.3
7	0, 0, 0.1	0, 0, 0.1	0.81, 1.0, 1.0
8	0.49, 0.81, 1.0	0.07, 0.27, 0.5	0.81, 1.0, 1.0
9	0.21, 0.45, 0.7	0, 0, 0.1	0.45, 0.7, 0.9
10	0.49, 0.81, 1.0	0.49, 0.81, 1.0	0.27, 0.5, 0.7
11	0, 0.09, 0.3	0.49, 0.81, 1.0	0, 0.1, 0.3
12	0.21, 0.45, 0.7	0.35, 0.63, 0.9	0.9, 0.3, 0.5
13	0, 0, 0.1	0.21, 0.45, 0.7	0.45, 0.7, 0.9
14	0.7, 0.27, 0.5	0.21, 0.45, 0.7	0.81, 1.0, 1.0
15	0.63, 0.9, 1.0	0.35, 0.63, 0.9	0, 0.1, 0.3
16	0, 0, 0.1	0.07, 0.27, 0.5	0.45, 0.7, 0.9
17	0.49, 0.81, 1.0	0, 0, 0.1	0.81, 1.0, 1.0
18	0.21, 0.45, 0.7	0, 0.09, 0.3	0.27, 0.5, 0.7
19	0.49, 0.81, 1.0	0.35, 0.63, 0.9	0.81, 1.0, 1.0
20	0, 0.09, 0.3	0.35, 0.63, 0.9	0.63, 0.9, 1.0
21	0.21, 0.45, 0.7	0.35, 0.63, 0.9	0, 0, 0.1
22	0, 0, 0.1	0.07, 0.27, 0.5	0.45, 0.7, 0.9
23	0.7, 0.27, 0.5	0.35, 0.63, 0.9	0.9, 0.3, 0.5
24	0.63, 0.9, 1.0	0.63, 0.9, 1.0	0, 0.1, 0.3
25	0, 0, 0.1	0.21, 0.45, 0.7	0.9, 0.3, 0.5
26	0.49, 0.81, 1.0	0.07, 0.27, 0.5	0, 0.1, 0.3
27	0.21, 0.45, 0.7	0.21, 0.45, 0.7	0, 0.1, 0.3

- The unification of fuzzy-MOORA, using the concepts of fuzzy set theory, was perceived to be a competent and acceptable effort in attaining the best parametric combination to confirm high productivity without compromising the quality. However, this might be limited to the selected range of process variables.

Table 12 Crisp values for weighted normalized fuzzy decision matrix

Alternative	Responses		
	F_c	R_a	VB
1	0.843	0.843	0.937
2	0.490	0.767	0.937
3	0.453	0.453	0.567
4	0.033	0.627	0.937
5	0.490	0.767	0.937
6	0.767	0.453	0.133
7	0.033	0.033	0.937
8	0.767	0.280	0.937
9	0.453	0.033	0.683
10	0.767	0.767	0.490
11	0.130	0.767	0.133
12	0.453	0.627	0.567
13	0.033	0.453	0.683
14	0.490	0.453	0.937
15	0.843	0.627	0.133
16	0.033	0.280	0.683
17	0.767	0.033	0.937
18	0.453	0.130	0.490
19	0.767	0.627	0.937
20	0.130	0.627	0.843
21	0.453	0.627	0.033
22	0.033	0.280	0.683
23	0.490	0.627	0.567
24	0.843	0.843	0.133
25	0.033	0.290	0.567
26	0.767	0.280	0.133
27	0.453	0.453	0.133

Table 13 Overall assessment value

Alternative	Responses			y_i	Rank
	F_c	R_a	VB		
1	0.843	0.843	0.937	2.623	1
2	0.49	0.767	0.937	2.194	3
3	0.453	0.453	0.567	1.473	15
4	0.033	0.627	0.937	1.597	14
5	0.49	0.767	0.937	2.194	3
6	0.767	0.453	0.133	1.353	16
7	0.033	0.033	0.937	1.003	24
8	0.767	0.28	0.937	1.984	6
9	0.453	0.033	0.683	1.169	18
10	0.767	0.767	0.49	2.024	5
11	0.13	0.767	0.133	1.03	23
12	0.453	0.627	0.567	1.647	11
13	0.033	0.453	0.683	1.169	18
14	0.49	0.453	0.937	1.88	7
15	0.843	0.627	0.133	1.603	12
16	0.033	0.28	0.683	0.996	25
17	0.767	0.033	0.937	1.737	9
18	0.453	0.13	0.49	1.073	21
19	0.767	0.627	0.937	2.331	2
20	0.13	0.627	0.843	1.6	13
21	0.453	0.627	0.033	1.113	20
22	0.033	0.28	0.683	0.996	25
23	0.49	0.627	0.567	1.684	10
24	0.843	0.843	0.133	1.819	8
25	0.033	0.29	0.567	0.89	27
26	0.767	0.28	0.133	1.18	17
27	0.453	0.453	0.133	1.039	22

References

1. Budinski KG (1991) Tribological properties of titanium alloys. Wear 151(2):203–217
2. Palraj S, Venkatachari G (2008) Effect of biofouling on corrosion behaviour of grade 2 titanium in Mandapam seawaters. Desalination 230(1):92–99
3. Zitter H, Plenk H (1987) The electrochemical behavior of metallic implant materials as an indicator of their biocompatibility. J Biomed Mater Res 21(7):881–896
4. Lautenschlager EP, Monaghan P (1993) Titanium and titanium alloys as dental materials. Int Dent J 43(3):245–253
5. Aziz-Kerrzo M, Conroy KG, Fenelon AM, Farrell ST, Breslin CB (2001) Electrochemical studies on the stability and corrosion resistance of titanium-based implant materials. Biomaterials 22(12):1531–1539
6. Ezugwu EO, Wang ZM (1998) Titanium alloys and their machinability—a review. J Mater Process Technol 68:262–274
7. Ezugwu EO, Bonney J, Yamane Y (2000) An overview of the machinability of aero-engine alloys. J Mater Process Technol 134:233–253
8. Narutaki N, Murakhoshi A (1983) Study on machining of titanium alloys. CIRP Ann Manuf Technol 32(1):65–69
9. Jawaid A, Che-Haron C, Abdullah A (1999) Tool wear characteristics in turning of titanium alloy Ti-6246. J Mater Process Technol 92:329–334
10. Leyens C, Peters M (2003) Titanium and titanium alloys. Wiley Online Library
11. Settineri L et al (2014) An evaluative approach to correlate machinability, microstructures, and material properties of gamma titanium aluminides. CIRP Ann Manuf Technol 63(1):57–60
12. Lalwani D, Mehta N, Jain P (2008) Experimental investigations of cutting parameters influence on cutting forces and surface roughness in finish hard turning of MDN250 steel. J Mater Process Technol 206(1):167–179
13. Aouici H et al (2012) Analysis of surface roughness and cutting force components in hard turning with CBN tool: prediction model and cutting conditions optimization. Measurement 45(3):344–353
14. Asiltürk I, Neşeli S (2012) Multi response optimisation of CNC turning parameters via Taguchi method-based response surface analysis. Measurement 45(4):785–794
15. Hashmi KH et al (2016) Optimization of process parameters for high speed machining of Ti-6Al-4 V using response surface methodology. Int J Adv Manuf Technol 85(5–8):1847–1856
16. Tebassi H et al (2017) On the modeling of surface roughness and cutting force when turning of Inconel 718 using artificial neural network and response surface methodology: accuracy and benefit. Period Polytech Mech Eng 61(1):1
17. Chandrasekaran M, Muralidhar M, Krishna CM, Dixit U (2010) Application of soft computing techniques in machining performance prediction and optimization: a literature review. Int J Adv Manuf Technol 46(5–8):445–464
18. Abburi N, Dixit U (2006) A knowledge-based system for the prediction of surface roughness in turning process. Robot Comput Integr Manuf 22(4):363–372
19. Basheer AC, Dabade UA, Joshi SS, Bhanuprasad V, Gadre V (2008) Modeling of surface roughness in precision machining of metal matrix composites using ANN. J Mater Process Technol 197(1):439–444
20. Mathonsi T, Laubscher RF, Gupta K (2018) Investigation on high speed machining of titanium grade 2 under MQL conditions. In: Proceedings of 16th international conference on manufacturing research 2018, Skovde (Sweden), Advances in manufacturing technology XXXII, pp 69–74. IOS Press
21. Miller T, Gupta K, Laubscher RF (2018) An experimental study on MQL assisted high speed machining of NiTi shape memory alloy. In: Proceedings of 16th international conference on manufacturing research 2018, Skovde (Sweden), Advances in manufacturing technology XXXII, pp 80–85. IOS Press

22. Mathonsi T, Laubscher RF, Gupta K (2018) On machinability of titanium grade 4 under minimum quantity lubrication assisted high speed machining. In: Conference of the South African Advanced Materials Initiative (CoSAAMI-2018). IOP conference series: materials science and engineering, vol 430, p 012013
23. Gupta K, Laubscher RF (2016) MQL assisted machining of grade-4 titanium. In: Proceedings of international conference on competitive manufacturing (COMA), Stellenbosch (South Africa), pp 211–217, 27–29 Jan 2016
24. Singaravel B, Selvaraj T (2015) Optimization of machining parameters in turning operation using combined TOPSIS and AHP method. Tehnicki Vjesnik 22(6):1475–1480
25. Sofuoglu MA, Orak S (2017) A novel hybrid multi criteria decision making model: application to turning operations. Int J Intell Syst Appl Eng 5(3):124–131
26. Khan A, Maity K. Selection of optimal machining parameters in turning of CP-Ti grade 2 using a hybrid optimization technique
27. Khan A, Maity K (2016) Parametric optimization of some non-conventional machining processes using MOORA method. Int J Eng Res Afr 20:19–40. Trans Tech Publications
28. Khan A, Maity K (2017) Application of MCDM-based TOPSIS method for the selection of optimal process parameter in turning of pure titanium. Benchmark Int J 24(7):2009–2021
29. Khan A, Maity K (2018) Application potential of combined fuzzy-TOPSIS approach in minimization of surface roughness, cutting force and tool wear during machining of CP-Ti grade II. Soft Comput, 1–12
30. Zadeh L (1965) Fuzzy sets. Inf Control 8(3):338–353

Application of Multi-objective Genetic Algorithm (MOGA) Optimization in Machining Processes

Nor Atiqah Zolpakar, Swati Singh Lodhi, Sunil Pathak and Mohita Anand Sharma

Abstract Multi-objectives Genetic Algorithm (MOGA) is one of many engineering optimization techniques, a guided random search method. It is suitable for solving multi-objective optimization related problems with the capability to explore the diverse regions of the solution space. Thus, it is possible to search a diverse set of solutions with more variables that can be optimized at one time. Solutions of MOGA are illustrated using the Pareto fronts. A Pareto optimal set is a set of solutions that are non-dominated solutions frontier. With the Pareto optimum set, the corresponding objective function's values in the objective space are called the Pareto front. The conventional methods for solving multi-objective problems consist of random searches, dynamic programming, and gradient methods whereas modern heuristic methods include cognitive paradigm as artificial neural networks, simulated annealing and Lagrangian approcehes. Some of these methods are managed in finding the optimum solution, but they have tendency to take longer time to converge so that need much computing time. Thus, by implementing MOGA approach that based on the natural biological evaluation principle will be used to tackle this kind of problem. In this chapter authors attempts to provide a brief review on current and past work on MOGA application in few of the most commonly used manufacturing/machining processes. This chapter will also highlights the advantages and limitations of MOGA as compared to conventional optimization techniques.

Keywords Design-of-experiment · Machining · Genetic algorithm · Optimization

---

N. A. Zolpakar · S. S. Lodhi · S. Pathak (✉)
Faculty of Engineering Technology, Universiti Malaysia Pahang, Lebuhraya Tun Razak, 26300 Gambang, Kuantan, Pahang Darul Makmur, Malaysia
e-mail: sunilpathak@ump.edu.my; sunilpathak87@gmail.com

M. A. Sharma
IMS UNISON University, Mussoorie Diversion Road Makkawala Greens, Dehradun, Uttarakhand 248009, India

K. Gupta and M. K. Gupta (eds.), *Optimization of Manufacturing Processes*, Springer Series in Advanced Manufacturing, https://doi.org/10.1007/978-3-030-19638-7_8

1 Introduction

In the manufacturing field, the ultimate objectives are to produce high quality product with minimum cost and time constrains. In manufacturing field, the common step applied to produce a product is machining [1]. In some cases, a single product needs to undergo different types of machining processes to come into its final shape, size, and form. The success of machining operation in terms of best combination of productivity, machinability, cost and sustainability can only be achieved when perform under optimum set of process parameters. To accomplish this objective, one of the consideration methods is optimization techniques. For machining optimization, there are two main methods which are conventional or classical technique (Design of Experiment (DOE), and Mathematical Iterative Search) and modern or advanced Technique (Meta-Heuristic Search and Problem Specific Heuristic Search). Figure 1 presents various optimization tools and techniques used in the past research for optimization of machining parameters. A review of past work based on implementation of conventional techniques such as machining theory, experiment investigation (parametric study), and DOE etc. has been reported in [2]. For this chapter, the focus is on optimization of machining process using Genetic Algorithm (GA).

Genetic algorithms are exceptionally well known heuristic techniques which have been effectively utilised to address optimization issues of machining. Genetic algorithm approves the consistency of the numerical model. For example, when company gets a large order, planner in the company needs to schedule and come out with a gantt chart for the particular product. In the Gantt chart, the information related machining and process is stated. The Gantt chart also represents the connection activities, time, and cost to be spent by production line. This step involves a lot of parameters such

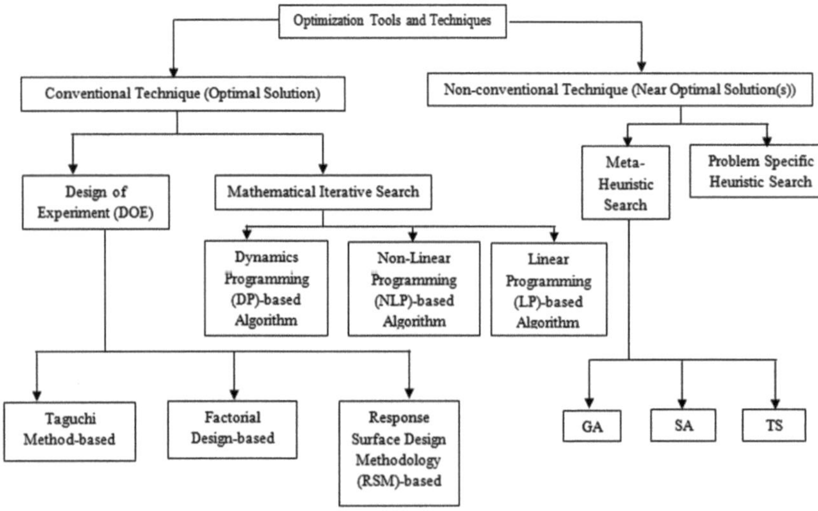

Fig. 1 Conventional (classical) and non-conventional (advanced) optimization techniques [3]

as machining time for every geometry, time for changing tools, surface roughness, and power consumtion. The function of GA is to find the combination of parameters to obtain a set of parameters that produces the optimum results. GA is one of the advanced techniques based on meta-heuristic search. One of the advantages of GA optimization including avoid converging into local minimum/maximum and instead this algorithm is able to find global minimum/maximum in the search space. Besides that, GA algorithm has capability to optimize more than one parameters in a single algorithm. This characteristic is important to apply in machining processes since machining process has a lot of parameters that need to be optimized such as spindle speed, feed rate, depth of cut, and axial angel of cutting etc.

Parameter selection is a critical part in optimizing the machining process in order to attain effective machining operation [4]. The selection of parameters typically based on the human judgement and experience. Due to that, most of the time, the selected process parameters does not provide an optimal result due to the fact that each of the parameters interrupt the process in getting optimum performance and quality. In fact, each machining parameter significantly affect other parameters as well. Based on that fact, a number of researchers used meta-heuristic search such as GA in their optimization (see Table 1). The following section will further discusses about application of GA optimization in machining process.

2 Genetic Algorithm (GA)

Genetic algorithm (GA) which was initially introduced by John Holland in 1975, is one of the classes for transformative algorithms that have been widely utilized in optimization problems. In spite of the fact that it is normal and conceivable to take care of issues with single target work, significant advantages of using GA are as follow:

- Real-life engineering problems usually demand for more than one objective functions and the GA is used to analyze various objective functions simultaneously.
- More than one parameter can be optimized.
- Optimization results are represented in the Pareto front form. It shows the combination of parameters with the values of objective function/s.
- Optimization results remain in the domain of the search area. Users have agility to define the size of the search area, and this avoids extreme results.

Today, due to the involvement of more complex engineering systems and processes, the optimum solutions are mainly trade-off based, where it is on the user description to select the appropriate and preferable decision criteria [37]. In GA optimization, if the algorithm has more than one objective functions, and one function is more important than the other, in that case the user needs to declare this by assigning weightage for every function. This new scheme of evaluating competing solutions without the necessity to determine relative importance weights, has given rise to multi-objective genetic algorithms (MOGA). In literature, MOGA has been

Table 1 Summary of past work based on GA optimization in machining process (2010–2018)

No.	Researcher	Process para meter	Machining process	Machining performance measure
1	Sekulic et al. [5]	Spindle speed, feed per tooth, axial depth, and radial depth	Ball-end milling	Surface roughness
2	Shukla and Singh [6]	Transverse speed, standoff distance, and mass flowrate	Abrasive wafer jet machining	Kerf top width and angle
3	Sangwan and Kant [7]	Cutting speed, feed, depth of cut	Turning	Energy consumption
4	Kumar et al. [8]	Cutting speed, feed rate, depth of cut, type of cutting tool	Turning	Surface finish
5	Kant and Sangwan [9]	Cutting speed, feed rate, depth of cut	Drilling milling	Surface roughness
6	Li et al. [10]	Speed, feed per tooth, width and depth of cut	Milling	Tool life, residual stress and surface roughness
7	Santos et al. [11]	Cutting speed, feed rate, depth of cut	Turning	Machining force, chip thickness ratio, and chip disposal
8	Manesh et al. [12]	Spindle speed, feed rate, axial depth of cut, and radial depth of cut	End milling	Surface roughness, MRR
9	Sahali and Serra [13]	Cutting speed, depth of cut	Turning	Production time
10	Sangwan et al. [14]	Cutting speed, depth of cut and feed rate	Turning	Surface roughness
11	Shivasheshadri et al. [15]	Speed and feed rate	Milling	Machining time
12	Agrawal and Varma [16]	Speed, feed	Milling	Surface roughness
13	Durairaja and Gowri [17]	Speed, feed, and depth of cut	Micro tuning	Surface roughness
14	Petkovic and Radovanovic [18]	Cutting speed and feed	Turning	Production cost
15	Selvam et al. [19]	Number of passes, cutting depth, spindle speed, and feed rate	Face milling	Surface roughness

(continued)

Table 1 (continued)

No.	Researcher	Process para meter	Machining process	Machining performance measure
16	Rai et al. [20]	Axial depth of cut, radial immersion, feed rate and spindle speed	Multi-tool milling	Machining time
17	Zeng et al. [21]	Rotate speed, speed and depth of cutting	N/A	Surface roughness
18	Gao et al. [22]	Bonding wear, feed per tooth and axial depth of cut	High speed machining	Cutting force, tool life
19	An et al. [23]	Speed, feed rate, depth of cut, and the number of passes	Multi-pass milling	Production cost
20	An [24]	Speed, feed rate and depth of cut	Multi-pass milling	Production cost
21	Kilickap et al. [25]	Cutting speed, feed rate, and cutting environment	Drilling	Surface roughness
22	Kuruvila and Ravindra [26]	Pulse-on and off duration, current, bed-speed and flushing rate	WEDM	Dimension error, surface roughness, volumetric MRR, production time
23	Ganesan et al. [27]	Depth of cut, cutting speed and cutting rate	Multi-pass turning	Production time
24	Xie and Guo [28]	Depth of cut, cutting speed and cutting rate	Multi-pass turning	Production cost
25	Zain et al. [29]	Cutting speed, feed rate, and radial rake angle	End milling	Surface roughness
26	Zain et al. [30]	Traverse speed, waterjet pressure, standoff distance, abrasive flow rate	Abrasive waterjet machining	Surface roughness
27	Zain et al. [31]	Cutting speed, feed rate and radial rake angle	End milling	Surface roughness
28	Zain et al. [32]	Radial rake angle, cutting speed and feed	End milling	Surface roughness

(continued)

Table 1 (continued)

No.	Researcher	Process para meter	Machining process	Machining performance measure
29	Sultana and Dhar [33]	Feed rate, pressure, flow rate and high pressure coolant	Turning	Chip reduction coefficient and surface roughness
30	Yongzhi et al. [34]	Axial depth-of-cut, radial depth-of-cut and helical angle	High speed milling	Cutting force, metal removal rate
31	Pasam et al. [35]	Ignition pulse current, short pulse duration, time between two pulses, servo speed, servo reference voltage, injection pressure, wire speed and wire tension	Wire electrical discharge machining	Surface roughness
32	Ansalam Raj and Narayanan Nambodiri [36]	Feed, speed rate, and depth of cut	NC milling	Surface roughness

reported superior compared to other classical algorithms [38]. In recent years, the GAs in machining application have been used by a number of researchers to find the optimal surface quality in various traditional and modern machining [5, 8, 9, 17]. Besides of surface roughness, many researchers applied GA optimization for minimize production cost and production time [13, 24, 27]. There are some variants of GA, some of the most commonly used in machining are as follows:

(a) ***Factual Coded Genetic Algorithm (FCGA)***: In FCGA, every gene signifies to a variable of the problem, and the extent of the chromosome is kept the same as the length of the response for the issue. In this way, FCGA can manage substantial areas without compromising with its accuracy as the binary execution. Moreover, FCGA has the ability with regards to the nearby tuning of the responses; it additionally permits integrating the domain knowledge in order to enhance the execution of Genetic Algorithm (GA).
(b) ***Binary coded Genetic Algorithm***: Binary coded Genetic Algorithm (BCGA) is a probabilistic search algorithm that iteratively changes a set (called as a population) of numerical items (typically settled length paired character strings), each associated with a fitness value, into another populace of posterity objects utilizing the Darwinian rule of regular choice and utilizing activities that are designed after normally happening genetic tasks, for example, hybrid (sexual recombination) and transformation. Following the model of development, they build up a population of individual, where every individual relates to a point in the hunt space. A target work is connected to every person to rate their wellness.

(c) ***Differential Evolution***: Differential Evolution (DE) tries to supplant the traditional hybrid and transformation plans of the genetic algorithm (GA) by elective differential administrators. The DE algorithm has as of late turned out to be very famous in the machine insight and computer science network. Much of the time, it has beaten the GA or the particle swarm enhancement (PSO). As in other developmental algorithms, two basic processes drive the advancement of a DE populace: the variety procedure, which empowers investigating the diverse districts of the inquiry space, and the determination process, which guarantees misuse of the obtained information about the wellness scene.
(d) ***Least Mean Square Algorithm***: Least mean squares (LMS) algorithms are utilized in versatile channels to discover the channel coefficients that identify with delivering the minimum mean squares of the blunder flag (difference between the desired and the actual signal). It is a stochastic inclination drop technique in which the channel is versatile in view of the blunder at the present time. The LMS algorithm can be actualized without squaring, averaging or separation and is a basic and effective process.
(e) ***Sawtooth Genetic Algorithm***: Various strategies have been produced to enhance the heartiness and computational proficiency of GAs. A straightforward GA utilizes a populace of consistent size and aides the development of an arrangement of haphazardly chose people through various ages that are liable to progressive determination, hybrid, and transformation, in view of the measurements of the age (standard GA). Population (data set) size is one of the principle parameters that influence the power and computational productivity of the GAs. Little populace sizes may result in untimely merging to non-ideal arrangements, while extensive populace sizes give a significant increment of computational exertion. A few strategies have been proposed in the writing that endeavors to build the decent variety of the populace and maintain a strategic distance from untimely merging.

2.1 GA Methodology

The GA algorithm start with randomly created initial population. Initial population is created by randomly form binary number. Every set of binary code that represent the solution is called chromosome. The length of the chromosome, *L*, is equal to the number of the bit in the string. There are $2^L - 1$ possible solution for selection and each solution is presented by *L*-bit binary code of chromosome, *C*. The optimization began with initialisation of a chromosome that contains the parameters to be optimized. A general representation is shown below:

$$C_k = [X_{k1},\ X_{k2}, \ldots X_{kn}]$$

$$C_k = [||110 \ldots 00||101 \ldots 1||001 \ldots 11||110 \ldots 11||]$$

X_1 X_2 X_3 X_4

Where X is represent the parameters that need to be optimized. From this initial population, a population that has a better representation of the strong species generated through selection process.

In GA, the selection process is based on the best individual performance on fitness function. The performance evaluation is depended on the fittest objective function. For minimization optimization problem, individual's chromosome with a smaller value of the fitness function will have higher possibilities selected for producing offspring. In the aforesaid explanation, the selection process is one in which the individuals that undergo genetic operations and come out with the offspring solutions. The selection has two primary objectives:

1. To choose the fittest individuals chromosome that can be directly copied for the next generation (elitism).
2. To give a chance to individual's chromosome with the fitness function that relatively bad value to partake in the process of the subsequent generations. To accomplish this, the observation of the global character of the search process is needed by not allowing a single individual dominate the population.

The selection process will create the intermediate population. The intermediate population is allowed to mate through cross-over and to modify through mutation and thus produce the next set of population. In the crossover operator, two solutions (parent) are chosen in the mating pool and at random point of string and some portion of the string are switched between the two solutions to create a new solution or offspring.

Parents	Offspring
0 1 1 0 | 0 1 1 1 | 0 1 0	0 1 1 1 | 0 1 1 1 | 0 1 0
0 0 1 1 | 0 1 1 1 | 0 0 0	0 0 1 0 | 0 1 1 1 | 0 0 0

Meanwhile, the mutation operator modifies a string locally to expectantly generate a better string. The bit-wise mutation process necessitates the construction of a random number for every bit. This procedure is repeated until the termination condition is reached [39]. Population is a collection of chromosomes that randomly initialized. The population get more fit with the search progress. The two operators that improve the population fitness are crossover and mutation. The flowchart of GA algorithm is shown in Fig. 2. The step-by-step procedure to apply GA in optimizing machining processes are listed as follow:

i. The selected parameters are encoded from real number to binary by binary encoding.
ii. A chromosome is performed by combination of a set of genes which this set is used to perform crossover and mutation.

Fig. 2 Flowchart of GA optimization

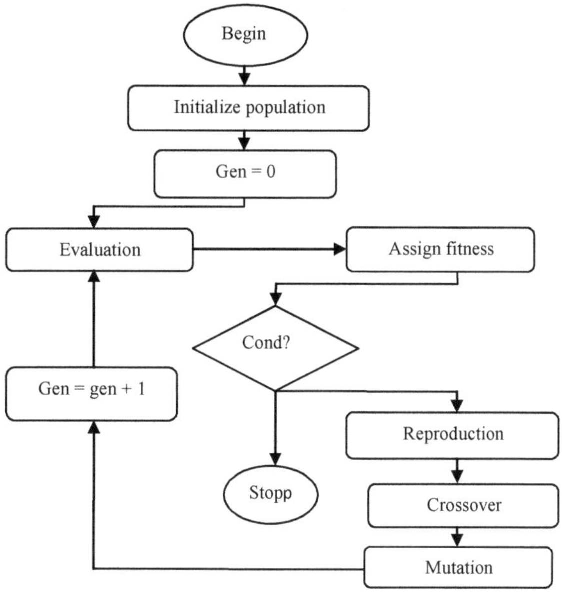

iii. Crossover operator will combine two chromosomes from population to form new chromosome that called offspring. The offspring chromosome expected to have better genes compared to the parent. As the crossover operator applied, the good chromosome will appear in the population and provide an overall good solution.

iv. Mutation is the process that applied after crossover operation. The mutation operator will apply random changes into a string of chromosome. The mutation process will help to overcame trapping at local minima.

v. The evaluation of chromosome is determinate by encoding from binary codes of chromosome to machining parameters values that can be used to estimate the machining performance.

vi. Objective function or fitness function is the function that needs to be maximize/minimize in the machining operation. This function must contain all the parameters that need to be optimized. The values of fitness function can be used as indication whether the parameters to be optimized or not.

vii. The iteration of the algorithm will continue until certain stopping criterion is made. One of the stopping criteria usually used is when value of fitness function of previous generation is less than 1×10^{-7} with the subsequent generation.

2.2 GA in Machining Applications

GA algorithm has the ability to optimize more than one parameter and more than one objective function simultaneously. This characteristic is very crucial in machining performance by optimizing several machining parameters to satisfy one or more objective functions. Moreover, in machining processes, machining conditions have an effect on diminishing the production cost and time and choosing the nature of the end product. To discover ideal cutting parameters amid a turning/milling/drilling/advanced machining procedure, the genetic algorithm has been successfully implemented. Process optimization needs to yield the least production time while thinking about innovative and material limitations. Target work is to decide the ideal machining parameters amid a machining procedure that limits the production time without disregarding any forced cutting imperatives. In the present work authors have tried to summarize few of the articles where MOGA has been used as optimization tool for machining application. From the review it was observed that majority of the work in machining was done to optimize the surface quality in terms of finish by optimizing machining conditions (almost 56%), production cost and time (almost 28%) and others such as cutting tool life span and energy consumption takes the stake of 16% as presented in Fig. 3.

Various researchers have done work using artificial intelligence for optimization of manufacturing and machining processes. This includes optimization of conventional machining such as turning, milling, cutting and drilling, and advanced machining techniques which includes electrochemical machining, electrical discharge machining, wire-electrical discharge machining and many more. The optimization has been used either to optimize the performance or improve the production cost and productivity. Aggarwal and Singh [2] have presented a detailed review on optimization of machining techniques using advanced optimization techniques including details of methodology and implementation of genetic algorithm (GA). Evolutionary algorithm and its comparison with various optimization techniques have been presented by Alberto et al. [39]. They have also developed new pareto rankings and compared them with the conventional methods.

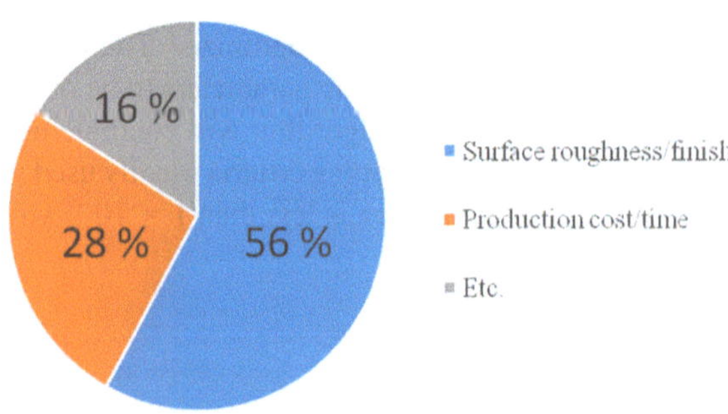

Fig. 3 Distribution of the research objectives in the previous study (2010–2018)

In the most recent work Sekulic et al. [5] have used response surface methodology (RSM), genetic algorithm (GA) optimization and grey wolf optimizer (GWO) algorithm for optimization of in ball end milling for prediction of surface roughness of hardened steel. They used predefined reduced-quadratic model as a benchmark model to develop GA and GWO algorithm. There results suggest 89.58% accuracy for GA model for training and testing data. Shukla and Singh [6] have used Taguchi method and Evolutionary optimization techniques in abrasive jet machining to optimize the transverse speed, stand-off distance and mass flow rate for attaining optimum values of kerf-top width and taper angle. They also used regression analysis to correlate the data of experimental findings. Sangwan and Kant [7, 14] have used integrated response surface methodology with genetic algorithm to optimize the energy efficiency in machining of AISI steel in turning, they also used GA to optimize the surface finish of the workpiece in turning operation. Sangwan and Kant [9] found experimental values and predicted results quite close with mean relative error is 4.11% showing fine accurateness in predicting the surface roughness values in ANN model joined with GA.

Kumar et al. [8] used GA to optimize the surface finish of the aluminum alloy composite. They praise the capabilities of GA in optimization of independent process parameters of machining methods. Multi-objective study on turning operation to find optimum cutting conditions for aluminium alloy using GA was done by Santos et al. [11]. Their study involves optimization of cutting speed, feed rate, and depth of cut on various inter related responses namely machining force, chip thickness ratio (CTR), and chip disposal. Durairaja and Gowri [17] had obtained the optimized cutting conditions for both surface roughness and tool wear by optimization of process parameters and statistical modeling using the multi objective genetic algorithm with valid experimental results. Petkovic and Radovanovic [18] obtained with minimal cost for the turning process, optimal parameters of machining (cutting speed and feed) were determined. Similar outcome was obtained during the use of GA checked by SQP (Sequential Quadratic Programming) algorithm and of machining cost, cutting speed and feed found with the GA.

According to Gao et al. [22], it was very essential to logically optimize cutting parameters prior to machining while the cutting force and tool wear have significantly reduced and cutting efficiency improved.

Training, testing and application subsequent to optimized 300 steps was adapted by Zeng et al. [21] resulting with the test error less than 2.6% with average relative error tended to saturation training was 4.0%.

Similarly, Sahali and Serra [13], Sultana and Dhar [33] and various other researchers have used GA as primary optimization tool to optimize the machining condition and responses in turning operations. The non-traditional algorithms were formulated by Ganesan et al. [27] where the optimal machining parameters for the continuous profile, GA and PSO have been employed. PSO produces better results with minimized time and Xie and Guo [28] have used GA to optimize the parameters in multi-pass turning for different materials, the complexity of optimization if multi-pass turning has been effectively eased by using GA.

Genetic Algorithm has been effectively used in optimization of milling parameters, many researchers have used to identify the optimum combination of parameters of milling using GA to obtain best results. Santos et al. [11] showed significant effect on the responses by the results of the input parameters acting both individually or in combination with each other. Li et al. [10] solved the multi-objective optimization problem by non-dominated sorting genetic algorithm-II (NSGA-II) and the Pareto-optimal solutions was obtained. The relative errors of surface roughness, tool life, and residual stress were less than 7, 5, and 5%, respectively after comparison of optimized results and experimental results.

Machining time was reduced by minimizing the negative effect to the part quality by Shivasheshadri et al. [15]. Initially required machining parameters (speed, feed and depth of cut) were given and 3D model was created which was undergone by five milling operations (facing, cornering, pocketing and two slot milling). Agrawal and Varma [16] proposed that GA can attain better-quality solutions to other metaheuristics to optimize the parameters of other machining processes (drilling and unconventional machining). By using RSM within the specified limits the optimal surface roughness value can be attained. The genetic algorithm (GA) model was trained and tested in MATLAB by Manesh et al. [12] to discover the best possible cutting parameters leading to least surface roughness (recommended 0.25 μm). Selvam et al. [19] used Taguchi technique that was fine-tuned with Genetic algorithm for finding Optimum machining parameter combination. The surface roughness evaluated through genetic algorithm it was 0.88 μm with 4.625% error from the predicted value and for Taguchi technique was 0.975 μm with 4.308% error from the predicted value. The different methods (integer programming, genetic algorithms and nonlinear programming) were used by An et al. [24] for obtaining optimal values of machining parameters. They match up the results from the literature and machining data handbook. Approximation algorithms used by An [23] developed the methods useful to optimize grinding and drilling type processes. The optimal cutting conditions were analyzed and obtained by Zain et al. [29] that yielded 0.138 μm as the minimum surface roughness value. The GA technique has reduced 27% of the least surface roughness value of the experimental sample data, 26% of regression modeling and 50% of response surface methodology technique. The R_a value was compared by Zain et al. [31] at about 26.8% to the experimental, 25.7% regression, 26.1% ANN and 49.8% response surface method in the reduced ANN–GA integration system. It was as well establish in comparison to the conventional GA result that integrated ANN–GA reduced the mean R_u value at about 0.61% and the number of iterations in searching for the optimal result at about 23.9%. Zain et al. [30] proposed that by means of the integrated SA–GA, the time for penetrating the optimal solution can be made quicker. A full-factorial experimental design and multi-linear regression technology were used by Yongzhi et al. [34] for developing the predictive model of surface roughness, for obtaining minimum cutting force and reasonably good metal removal rate it was possible to select optimum axial depth-of-cut, radial depth-of-cut and helical angle. Rai et al. [20] also have used multi-objective genetic algorithm (MOGA) for optimization of parameters of milling namely Speed, feed rate, depth of cut, radial rake angle and the number of passes on surface quality of different materials. Ansalam and Nambodiri [36] used MOGA for optimization of surface roughness in numerical control milling machines, they considered effects of feed, speed rate, and depth of cut for multi-objective optimization techniques. MOGA has also been used to optimize advanced machining processes such as abrasive jet machining (AJM), EDM, WEDM, ECM, ECH and PECH [40, 41] etc. In AJM traverse speed, waterjet pressure, standoff distance, abrasive flow rate was considered as most frequently used input parameter while surface roughness has been selected as response [32]. Kuruvila and Ravindra [26] and Pasam et al. [35] have used MOGA to analyze and optimize the pulse current, pulse duration, pulse interval, servo speed, servo voltage, wire speed and wire tension during wire-EDM. Simultaneous optimization of such variety of parameters were possible at same time due to the use of genetic algorithm.

The afore discussed literature review is summarized in Table 1.

3 Conclusion

Multi-objective genetic algorithm technique has widely been employed for optimization of machining parameters to secure the best possible values of various machinability indicators such as surface roughness, material removal rate, and surface integrity etc.

GA optimization in optimizing machining parameters showed positive results based on literature review. Based on the review, most of the researchers used single-objective GA in their optimization scheme. By doing this, the other outcomes of machining is ignored even thought the same parameters will contribute to that outcome. Thus, as suggestion for the future researchers, Multi-Objective GA (MOGA) can be implemented in optimizing machining process without neglecting other properties. For now, the main concern is surface roughness and production cost, by making one of this as objective function, another function need to be sacrificed. To obtain maximum quality of surface roughness, production cost gets higher. Due to that fact, implementation of MOGA techniques will balance out the objective function and produces high quality surface roughness within the cost limitation.

In terms of machining, every different setup of machining with different type of workpiece, type of machining work and coolant used, and other parameters will provide unique solution set of optimization for particular setup when apply GA optimization. This showed that GA optimization is capable to provide technologist the required parameters for optimum machining processes.

References

1. Gupta K, Gupta MK (2019) Developments in non-conventional machining for sustainable production: a state of art review. Proc Inst Mech Eng C J Mech Eng. https://doi.org/10.1177/0954406218811982
2. Aggarwal A, Singh H (2005) Optimization of machining technique—a retrospective and literature review. Sadhana-Acad Proc Eng Sci 30: 699–711
3. Mukherjee I, Ray PK (2006) A review of optimization techniques in metal cutting processes. Comput Ind Eng 50:15–34
4. Magabe R, Sharma N, Gupta K, Davim JP (2019) Modeling and optimization of wire-EDM parameters for machining of $Ni_{55.8}$-Ti shape memory alloy using hybrid approach of Taguchi and NSGA-II. Int J Adv Manuf Technol. https://doi.org/10.1007/s00170-019-03287-z
5. Sekulic MA, Pejic VB, Brezocnik MC, Gostimirović MA, Hadzistevic MA (2018) Prediction of surface roughness in the ball-end milling process using response surface methodology, genetic algorithm, and grey wolf optimizer algorithm. Adv Prod Eng Manag 13:18–30
6. Shukla R, Singh D (2016) Experimentation investigation of abrasive water jet machining parameters using Taguchi and evolutionary optimization technique. Swarm Evol Comput 32:167–183
7. Sangwan KS, Kant G (2017) Optimization of machining parameters for improving energy efficiency using integrated response surface methodology and genetic algorithm approach. Procedia CIRP 61:517–522
8. Kumar KP, Manikandan K, Nandhakumar M, Rajendran KL (2015) Optimisation of machining parameters in aluminium alloy composite using genetic algorithm. Int J Sci Eng 1(1)
9. Kant G, Sangwanb KS (2015) Predictive modelling and optimization of machining parameters to minimize surface roughness using artificial neural network coupled with genetic algorithm. Procedia CIRP 31:453–458
10. Li J, Yang X, Ren C, Chen G, Wang Y (2015) Multiobjective optimization of cutting parameters in Ti-6Al-4V milling process using nondominated sorting genetic algorithm-II. Int J Adv Manuf Technol 76:941–953
11. Santos MC Jr, Machado MR, Barrozo MAS, Jackson MJ, Ezugwu EO (2015) Multi-objective optimization of cutting conditions when turning aluminum alloys (1350-O and 7075-T6 grades) using genetic algorithm. Int J Adv Manuf Technol 76:1123–1138
12. Mahesh G, Muthu S, Devadasan SR (2014) Prediction of surface roughness of end milling operation using genetic algorithm. Int J Adv Manuf Technol 77:369–381
13. Sahali MA, Belaidi I, Serra R (2015) Efficient genetic algorithm for multi-objective robust optimization of machining parameters with taking into account uncertainties. Int J Adv Manuf Technol 77:677–688
14. Sangwan KS, Saxenaa S, Kanta G (2015) Optimization of machining parameters to minimize surface roughness using integrated ANN-GA approach. Procedia CIRP 29:305–310
15. Shivasheshadri M, Arunadevi M, Prakash PS. Simulation approach and optimization of machining parameters in CNC milling machine using genetic algorithm. Int J Eng Technol 1(10):1–10
16. Agarwal A, Varma SN (2015) Optimization of machining parameters for milling operations using a genetic algorithm approach. Int J Eng Technol Res 3(1)
17. Durairaja M, Gowri S (2013) Parametric optimization for improved tool life and surface finish in micro turning using genetic algorithm. Procedia Eng 64:878–887
18. Petkovic D, Radovanovic M (2013) Using genetic algorithms for optimization of turning machining process. J Eng Stud Res 19(1):47–55
19. Selvam MD, Shaik Dawood AK, Karuppusami G (2012) Optimization of machining parameters for face milling operation in a vertical CNC milling machine using genetic algorithm. Eng Sci Technol Int J (ESTIJ) 2(4):2250–3498
20. Rai JK, Brand D, Slama M, Xirouchakis P (2011) Optimal selection of cutting parameters in multi-tool milling operation using a genetic algorithm. Int J Prod Res 49(10):3045–3068
21. Zeng HY, Qiang EJ, Yang XP, Li HM (2011) Soft-sensing model on the roughness of machining surface under the numerical control and its application. Appl Mech Mater 48–49:1077–1085

22. Gao DQ, Li ZY, Mao ZY (2011) Study of high speed machining parameters on nickel-based alloy GH2132. Adv Mater Res
23. An I, Feng I, Lu C (2011) Cutting parameters optimization for multi-pass milling operations by genetic algorithms. Adv Mater Res 160–162:1738–1743
24. An I (2011) Optimal selection of machining parameters for multi-pass turning operations. Adv Mater Res 156–157:956–960
25. Kilickap E, Huseyinoglu M, Yardimeden A (2011) Optimization of drilling parameters on surface roughness in drilling of AISI 1045 using response surface methodology and genetic algorithm. Int J Adv Manuf Technol 52:79–88
26. Kuruvila N, Ravindra HV (2011) Parametric influence and optimization of wire EDM of hot die steel. Mach Sci Technol 59:142–145
27. Ganesan H, Mohankumar G, Ganesan K, Ramesh Kumar K (2011) Optimization of machining parameters in turning process using genetic algorithm and particle swarm optimization with experiment verification. Int J Eng Sci Technol (IJEST) 3:1091–1102
28. Xie S, Guo Y (2011) Intelligent selection of machining parameters in multi-pass turning using a GA-based approach. J Comput Inf Syst 7(5):1714–1721
29. Zain AM, Haron H, Sharif S (2010) Application of GA to optimize cutting conditions for minimizing surface roughness in end milling machining process. Expert Syst Appl 37:4650–4659
30. Zain AM, Haron H, Sharif S (2011) Integration of simulated annealing and genetic algorithm to estimate optimal solutions for minimizing surface roughness in end milling Ti-6Al-4V. Int J Comput Integr Manuf 24(6):574–592
31. Zain AM, Haron H, Sharif S (2012) Integrated ANN-GA for estimating the minimum value for machining performance. Int J Prod Res 50(1):191–213
32. Zain AM, Haron H, Sharif S (2011) Estimation of the minimum machining performance in the abrasive waterjet machining using integrated ANN-SA. Expert Syst Appl 38:8316–8326
33. Sultana I, Dhar NR (2010) GA based multi-objective optimization of the predicted models of cutting temperature, chip reduction co-efficient and surface roughness in turning AISI 4320 steel by uncoated carbide insert under HPC condition. Paper presented at the proceedings of 2010 international conference on mechanical, industrial, and manufacturing technologist, MIMT, 2010, pp 161–167
34. Yongzhi P, Jun Z, Xiuli F, Xing A (2010) Optimization of surface roughness based on multi-linear regression model and genetic algorithm. Adv Mater Res 97–101:3050–3054
35. Pasam VK, Battula SB, Valli PM, Swapma M (2010) Optimizing surface finish in WEDM using Taguchi parameter design method. J Braz Soc Mech Sci Eng 32(2):107–113
36. Ansalam Raj TG, Namboothiri VN (2010) An improved genetic algorithm for the prediction of surface finish in dry turning of SS 420 materials. Int Adv Manuf Technol 47:313–324
37. Zolpakar NA, Ghazali NM, Hassan El-Fawal M (2016) Performance analysis of the standing wave thermoacoustic refrigerator, review. Renew Sust Energ Rev 54:626–634
38. Deb K (2001) Multi-objective optimization using evolutionary algorithm. Wiley, London
39. Alberto I, Azcarate C, Mallor F, Mateo PM (2003) Multiobjective evolutionary algorithms. Pareto rankings. Monogfias del Senim. Matem. Gracia de Galdeano. 27:27–35
40. Pathak S, Jain NK, Palani IA (2016) Investigations on surface quality, surface integrity and specific energy consumption in finishing of straight bevel gears by PECH process. Int J Adv Manuf Technol 85 (9–12):2207–2222
41. Pathak S, Jain NK, Palani IA. (2014) On use of pulsed-electrochemical honing to improve micro-geometry of bevel gears. Mater Manufact Process 29 (11–12):1461–1469

Optimization in Manufacturing Systems Using Evolutionary Techniques

Ravi Shankar Rai and Vivek Bajpai

Abstract This chapter introduces various types of manufacturing systems and different types of traditional and modern optimization techniques. Chapter briefs about evolutionary techniques such as Particle swarm optimization and Genetic algorithm to optimize the various kind of manufacturing system with an objective to overcome the limitations of traditional optimization techniques and to enhance the optimality of objective function. Besides that, customary methodology is to utilize an ordinary least squares relapse investigation for building up the machinability models. In recent decade, the utilization of evolutionary calculation techniques, or additionally called the Genetic strategies, in light of impersonation of Darwinian characteristic choice has turned out to be across the board. This is because of truth that numerous frameworks are excessively complex, making it impossible to be effectively enhanced by the utilization of traditional deterministic calculations. Despite what might be expected, the evolutionary algorithms (EA) include probabilistic tasks. The current chapter also presents brief details about stepwise procedure of implementation of genetic algorithm and particle swarm optimization to solve various problems associate with manufacturing systems.

Keywords Evolutionary · Genetic algorithm · Manufacturing · Optimization · Particle swarm

1 Introduction

In order to sustain in today's era of fluctuating and fierce market, manufacturing systems are to be flexible, efficient and productive. Such requirements can be achieve by following the principle of optimization which is known as the procedure of finding the fittest solution out of the numerous solutions. Therefore optimization is essential for making decisions in manufacturing system [1]. Mathematical techniques provide

R. S. Rai · V. Bajpai (✉)
Department of Mechanical Engineering, Indian Institute of Technology (ISM) Dhanbad, Dhanbad, India
e-mail: vivek@iitism.ac.in

K. Gupta and M. K. Gupta (eds.), *Optimization of Manufacturing Processes*, Springer Series in Advanced Manufacturing, https://doi.org/10.1007/978-3-030-19638-7_9

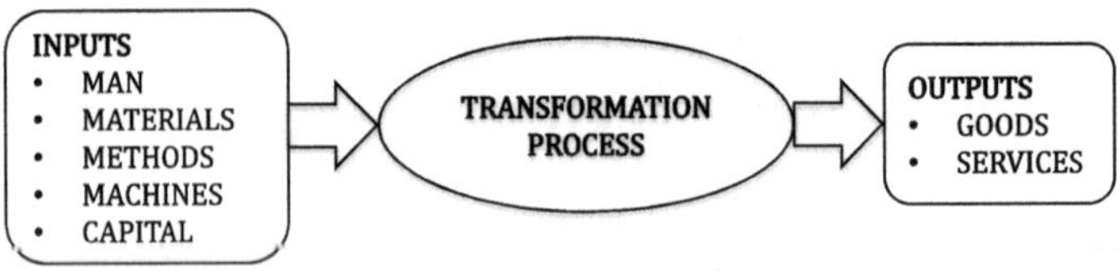

Fig. 1 Transformation process in manufacturing

foundation for solving problems having multiple variables and physical problem can be modeled in mathematical equations. The problems can be modeled in such a way that above requirement can fulfill and such function is known as objective function and in case of optimization these functions have to be either maximized or minimized. Objective functions can be expressed as a function of independent variables known as decision variables [2]. Every optimization problem must specify the range of decision variables which is called as geometric constraints. Several optimization techniques are developed over a few decades for optimizing different factors of manufacturing system, but evolutionary techniques have edge of solving problems in an effective manner. Operations segment is the key component of a manufacturing firm. Operations are defined as the process happening in the system to get work done. This consists of service operations and manufacturing operations. The process deals with the conversion of some raw materials into a useful product or service. Figure 1 illustrates the transformation process. Researchers are focusing on to optimize every operation of manufacturing systems that affect their performance.

2 Manufacturing System

Manufacturing system is defined as the collection or arrangement of different operations related to produce desired component. It consists of relevant infrastructure and machineries for performing and arranging those processes. Manufacturing system must be functionally efficient so that it can accommodate or adjust itself under any circumstances. Usually the occurrence of critical conditions or disturbances can counter by controlling the inputs or the system [3]. A generalized definition of manufacturing system is illustrated in Fig. 2.

2.1 Classification of Manufacturing Arrangements

Manufacturing arrangements can be categorized in terms of physical and structural features. As per the physical features, traditional manufacturing systems are categorized in four kinds such as Job based arrangement, Flow arrangement, Project arrangement and Continuous arrangement. There are some major categories of non-

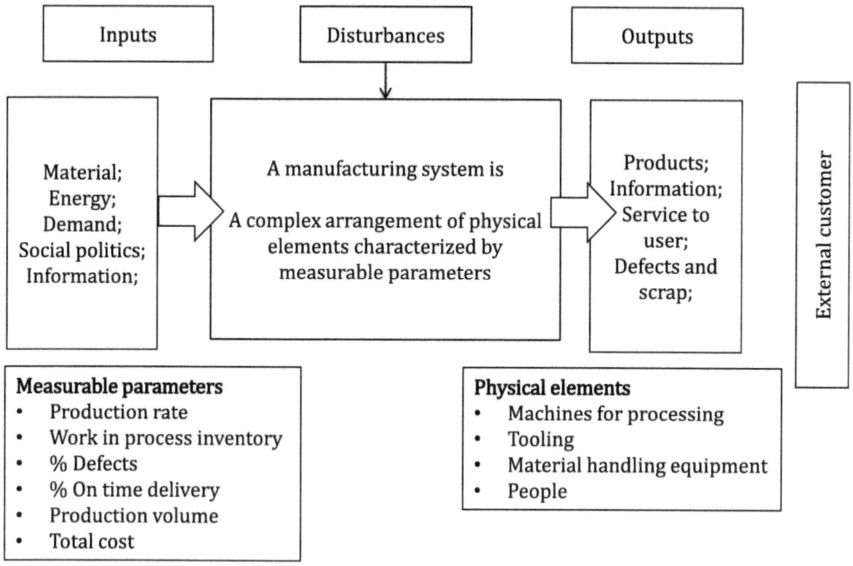

Fig. 2 Generalized frame for a manufacturing system [4]

traditional manufacturing are Dedicated manufacturing system (DMS), Cellular manufacturing system (CMS), Flexible manufacturing system (FMS), Reconfigurable manufacturing system (RMS) and Focused flexible manufacturing system (FFMS). In Job based arrangement assortments of items are made in little part sizes to an explicit client arrange. To play out a wide assortment of manufacturing forms, general purpose machines are needed. A group of skilled labors are used for performing various tasks. Machines are grouped as per the processes used for manufacturing of an item or product [5]. Flow arrangement is based "Product oriented layout" consisting of material flow line due to which arrangement can achieve high production rate. This arrangement uses "Special purpose machines" and designed to manufacture the specific product or family. Automated assembly lines and Television manufacturing factories are examples of such system [5]. Project shop systems are based on fixed position of product due to robust size and weight. The man, machines and materials have brought to site for fabrication. Such shop is named as fixed position shop [5]. The classical system which allows physical flow of product and it named as flow production while considering production of complex products like bottling process or assembling work like TVs. Nonetheless, this is definitely not a ceaseless procedure, however large quantity stream lines is named as continuous layout format [5].

2.2 Modern Manufacturing System

This system must have the capacity to adjust to sudden interior and exterior changes. An assortment of designed models and control procedures has been produced over the past two decades which is based upon the theory and tools of computer technology and management science. Under the recent industrial era, manufacturing organizations are confronting drastic variations driving to enhance their standards in product design development and execution. High adaptability, vigorous demand in market, expanding customization, astounding items, adaptable bunches and less life cycle are prominent elements driving the change from the classical system to the supposed Next Generation Manufacturing Systems (NGMSs) [6]. To cope up the bar of the present arrangements, advanced systems have to achieve great flexibility, AI-features and ease to reconfigure to resolve vigorous market conditions [7].

During the 1900s, there was a significant transformation in manufacturing sector. The principle cause was execution by Henry Ford arrangement of large scale manufacturing and devoted production arrangements. Model T was created in 1907. Model T has a specific measure of reputation, on the grounds that at one phase Henry Ford should had quoted for his brainchild "you can have any color, so long as it's black" [8]. The origination of large scale manufacturing presented by Ford, adds to advancement of DMSs which for the most part show up in two structures [9] such as Continuous DMS and Intermittent DMS. Continuous DMS will work to manufacture items in high orders and no explicit orders. Sales forecasting is an important work in this system but in intermittent framework, the products are fabricated uniquely to satisfy demands prepared by clients instead of for stock. This system runs on irregular stream of material. But CMS is a hybridized framework for connecting the upsides of both flow lines and job arrangements. A CMS is made out of "linked cells". Each cell of CMS is composed of flow shop arrangement of workstations [10]. This system allows modifications of machines, retooling and rearrangement inside equivalent "part family". Amidst the 1960s, demands of competitive market make organizations to stand up for advancements in production orientations. To resolve market challenges, flexible manufacturing system was evolved [11]. A cell of computerized numeric controlled machines operated by a common control unit will form a FMS. System has high operational flexibility because numerous product features can be manufactured with fast delivery [12]. Similarly RMS was concocted in 1999 in the "Engineering Research Center for Reconfigurable Manufacturing Systems (ERC/RMS) at the University of Michigan College of Engineering" [13]. Supreme objective of the RMS was outlined by the comment "Exactly the capacity and functionality needed, exactly when needed". The second advanced origination of production frameworks configuration is an origination of FFMS. These systems speak to likewise a focused response to adapt with requirement of customization and they ensure the ideal exchange-off among efficiency and adaptability [14].

2.3 Potential Requirement of Optimization in Manufacturing Systems

In the present economic condition, organizations are confronting complex difficulties caused by unstable markets, specific items, less life cycles, and worldwide competitiveness [15]. Manufacturing frameworks in quest for expense and time decrease without diminishing quality and adaptability are winding up increasingly perplexing. The comprehension and command over difficulties of system is exceptionally fundamental on the grounds that the non-linear conduct of production frameworks will definitely allow the system to be more gainful and prescient [16]. In this manner the system must be advanced dimension of execution in every part of the creation to satisfy the optimized necessities. Competitive era of economy imposed the system to have superior performance at least possible cost. In this manner, specific consideration must be taken to the choice of qualities for the distinctive elements which impact performance and expenses [17]. Components can be with respect to the arrangement of the physical framework (e.g. various equipment, logistics and storage concerns) or the executives parameters (e.g. storage strategies, dispatching variables, amount of Kanban). This can be tended to by optimizing a model, that is, to precisely pick the estimations of the n factors, X_i, of a vector $X = (X_1, X_2, \ldots, X_n)$, where the X_i factors can opt data from an element of the real set (e.g., velocity of AGV), an element of the integer set (e.g., number of spots in a stockroom) or in any usual set E (e.g., decision among various dispatching rules). In this manner to satisfy the necessities of the item or conquer the issues of the production framework, the framework needs optimum amount of operational and administrative resources for efficient working of the firms.

2.4 Approaches for Modeling of Manufacturing Systems

A model is defined as an exact portrayal of a framework. A precise model of a framework enables investigator to draw deductions about the framework under investigation without exploring different avenues regarding the real framework. When all is said and done, the contributions to a quantitative model are of two sorts: parameters (non-controllable factors) and choice factors (controllable factors). Design considers different set of elements of decision parameters. The refinement between decision factors and parameters were always not clear. Such as set-up time on a specific machine may represents parameters of one model but for other model it can be taken as decision variable [18]. The yields from a model of a production framework could incorporate performance estimates. The essential thought of model experimentation is to decide the qualities for the decision factors with the end goal that the performance factors are optimum.

This is a troublesome procedure for various reasons like regularly there are many clashing execution measures to be taken amid experimentation. Subsequently, investigator must worry about a multi-objective optimization in which contracts can made between different execution measures [19]. Next difficult element of model experimentation is the way that few of the decision factors might be number in nature. Mostly traditional optimization methods consider that taken problems have single local optima. Henceforth, every optimization technique has multiple local optimum values whose functional relations are not accessible. Approaches of modeling are relies upon following factors:

- Model definition
- Model formulation
- Model development
- Model validation
- Model evaluation.

To meet demand of the various optimization issues, following prominent modeling techniques were developed:

i. Linear programming,
ii. Nonlinear programming,
iii. Classical optimization techniques,
iv. Integer programming,
v. Dynamic programming,
vi. Stochastic programming,
vii. Geometric programming,
viii. Evolutionary algorithms, etc.

3 Optimization in Manufacturing Industries

The most simplified definition of optimization is "doing the most with the least". The procedure of calculating most favorable value is called as optimization [20]. The motivation behind optimization is to accomplish the "best" structure with respect to an arrangement of organized criteria. These incorporate maximizing elements like profitability, quality, life span, productivity, and usage [21]. In production system optimization is defined as the control of calculating the best option among a set, with in an explicit rule in the production condition. Optimization incorporates real problems and finds solution from model [22].

The purpose of manufacturing of a product is to deliver items satisfying intended functions, qualities, performances and attributes [23]. At each level of system optimization can be used and for that objective function with constraints must be formulated in each case. Linear programming model will be form from general manufacturing system to find out the optimal parameters to maximize the gain [24]. In subtleties let b the arrangement of assets of the manufacturing framework to be changed in item amounts x through the innovative modalities A. A is the innovative lattice and its nonexclusive component A_{ij} characterizes the asset of sort i expected to deliver j. Every row of the innovative lattice characterizes the amount of asset required for each unique item. Each column characterizes an explicit item, specifically the amounts of the diverse assets which must be utilized to create a unit of the item. The multiplication of an item column of A by the vector x gives the amount of asset that must be utilized to deliver the predetermined item and that must be not exactly accessible asset (pronounced in b). The target of the issue is to augment the benefit z. Therefore the issue can be planned in the accompanying way as mentioned in equation (i):

$$\mathrm{Max}\,Z = cx \tag{1}$$

Subject to, $Ax \leq b$
$x \geq 0$

So as to optimize manufacturing system, it is basic to plan items so as to permit successful optimization. The strong relation between design, manufacturing, and appropriation is an essential component for all encompassing optimization in production.

3.1 Traditional Optimization

The traditional optimization procedures are helpful in finding the optima point of a function and maxima or minima values. In this category utilizes the concepts of differential calculus to obtain optimal value, so called as analytical methods. These techniques assume that functions have double differentiability with respect to design parameters and their derivatives have continuous nature. These techniques have narrow utility as few problems contain discontinuous and not differentiable functions. But still these methods provide foundation for producing advance techniques to solve real world issues.

Three important categories of issues can be solved by the traditional optimization methods:

i. Single variable functions
ii. Multivariable functions without constraints
iii. Multivariable functions with both constraints (equality and inequality).

Conventional optimization methods initiate from randomly selected initial solution then propagates to optima point iteratively. Direction of search and step size is two prime factors to be selected by optimization algorithms. Large numbers of classical methods are in existence and they classified into two major category, namely direct search and gradient-based methods [25]. Direct search methods apply only function parameters at various points to search and never use partial derivatives of the functions that's why called as non-gradient techniques. But gradient based methods implement differential calculus on objective functions and constraints to obtain optimal result. In general, the techniques of optimization which need gradient values are taken more effective [26]. Some gradient based methods are steepest descent method, Newton's method, conjugate gradient method and variable-metric method. The most widely used direct search methods are Hooke-Jeeves method, Powell's conjugate direction method.

3.1.1 Disadvantages of Traditional Optimization

Numerous challenges like multi-methodology, differentiability and dimensionality are related with the optimization of large-scale issues. Conventional strategies namely steepest decent, dynamic programing and linear programing for the most part neglect to take care of such substantial issues particularly with nonlinear target functions. The greater part of the customary methods requires gradient data and consequently it is absurd to expect to explain non-differentiable issues with the assistance of such conventional procedures. Additionally, such procedures regularly fail to take care of optimization issues that have numerous local optima. To defeat these issues, researchers need to grow all the more incredible optimization methods and over the 30 years, considerable research has been proceeding to discover new techniques to effectively solve such issues. Classical optimization tools have the following disadvantages [27]:

1. Obtained solutions are reliant on the randomly selected initial point. Chance of calculated solution to be global optima is uncertain.
2. Discontinues function based optimization issues cannot be handled utilizing the gradient-based strategies. Additionally, the results of gradient techniques may stall out at local optima.
3. There exists an assortment of optimization issues. A specific conventional optimization strategy might be appropriate for taking care of just a single kind of issue. Along these lines, there is no flexible optimization strategy, which can apply to tackle an assortment of issues.
4. The convergence to an optima point relies upon the picked optima point.
5. Most of the algorithms cannot find suboptimal solution.
6. Algorithms are not productive in dealing with issues having discrete factors.
7. Algorithms cannot be productively utilized on parallel machine.

3.2 Advanced Optimization Techniques

To beat the disadvantages of the customary methods, analysts created advanced systems to tackle the optimization issues. The majority of the cutting edge optimization calculation depends on populace and natural or transformative hereditary qualities. People, have a characteristic propensity to pursue the manner in which the nature has tackled complex optimization issues, at whatever point we neglect to fathom them utilizing conventional improvement techniques. Some common procedures, like organic, physical procedures and so on are displayed artificially to create optimization tools for taking care of the issues.

A portion of the notable populace based procedures created over the 30 years are: Particle Swarm Optimization (PSO) [28] which relies on "the principle of foraging behavior of the swarm of birds"; Genetic Algorithms (GA) [29] which is based on "Darwinian theory of the survival-of-the-fittest and the theory of evolution of the living things"; Differential Evolution (DE) [30] which resembles with GA but have distinguished selection and the crossover; Ant Colony Optimization (ACO) [31] which relies on "the principle of foraging behavior of the ant for the food"; Artificial Immune Algorithms (AIA) [32] which is based on "the principle of immune system of the human being"; Artificial Bee Colony (ABC) [33] and so on. Numerous engineering problems can be solved by these algorithms and found effective to obtain solutions of explicit sort of issues.

4 Evolutionary Algorithms

Evolutionary calculation is the investigation of computational frameworks which utilizes thoughts from nature's evolution and adaptation. Numerous evolutionary calculation procedures get their thoughts and motivations from atomic advancement, populace hereditary qualities, immunology, and so forth. A portion of the phrasing utilized in evolutionary calculation has been acquired from these areas to mirror their associations like as genetic algorithms, mutation, crossover, phenotypes and species. From a regular perspective, an EA is a calculation that simulates—at some dimension of reflection a Darwin's evolutionary framework. To be more explicit, a standard EA incorporates:

1. At least one populaces of people viewing for constrained assets.
2. These populaces vary progressively because of the birth and death of people.
3. An idea of fitness which mirrors the capacity of a person to endure and reproduce.
4. An idea of modificational proliferation: posterity nearly looks like their folks, yet are not indistinguishable.

More or less, the Darwinian theory of evolution proposed that, by and large, species enhance their fitness over ages (i.e., their ability of adjusting to the earth).

4.1 Principle of Evolutionary Algorithm

These algorithms are based on stochastic way to search. EAs have two conspicuous highlights which separate themselves from other techniques. Initially, they are based on populace. Second, there is interchanges and data transfer amidst individuals in a populace. EAs for the most part continue on a basic level as indicated by the plan represented in Fig. 3.

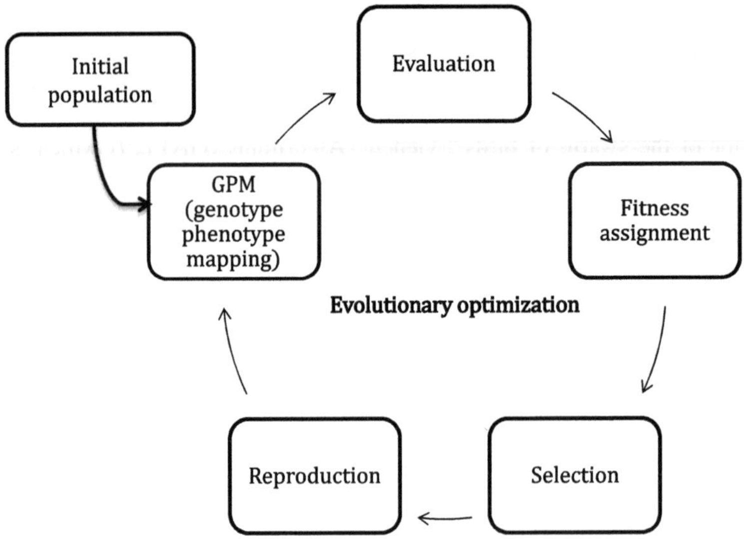

Fig. 3 The basic cycle of evolutionary algorithm [34]

Stepwise procedure can be illustrated as follows:

i. In first generation, a populace of n > 0 people is made. Generally, these people have arbitrary genotypes yet here and there, the underlying populace is seeded with great candidate solution either recently known or made by some different techniques.
ii. The genotypes, i.e., the values in the search region, are then meant phenotypes. For the situation that seeks tasks specifically deal with the solution information structures, this genotype-phenotype mapping is called as identity mapping.
iii. The estimations of the target function are then assessed for every applicant solution in the populace. This assessment may join confounded simulations and computations.
iv. In the objective function, applicability of various highlights of the candidate arrangements has been resolved. On the off chances that there is in excess of one objective function, constraints, or other applicability factor, at that point a scalar fitness point is allotted to every one of them.
v. An ensuing determination process sifts through the candidate arrangements with least fitness and permits those with great fitness to join in mating pool with larger likelihood.
vi. In the reproduction stage, offspring are gotten from the genotypes of the chose people by applying the search tasks. There are typically two distinctive reproduction tasks: mutation, which alters one genotype, and crossover, which joins two genotypes to another one.

If the final measure is satisfied, the advancement stops here. Else the evolutionary cycle proceeds with coming generation at point 2.

4.2 Optimization in Manufacturing Using Evolutionary Techniques

Bulgak et al. [35] studied the buffer improvement problems related to asynchronous systems. Choosing applicable buffer sizes for the transportation of automatic production systems could be a difficult job that has to take for random variations in productivity by the individual stations additionally as for delay in transports which is associated with material transport system. GA is applied to the present case in a trial to increase the appliance domain of GAs to the random style optimization issues of producing systems. Supported the obtained results, they all over that these advanced random combinatorial engineering issues may be resolved at least time with better accuracy.

Ozcelik and Erzurumlu [36] suggested ANN and GA to reduce warpage of skinny shell plastic elements made-up by injection molding. Computer button base is considered as sample of skinny shell plastic element. ANN hybrid with GA effectively applied to enable the objectives to attain optimum level and GA considerably diminishes the warpage of primary model and outcomes were enhanced by 51%.

Cook et al. [37] developed NN-GA hybrid for optimization of process for fiberboard production shop. GA was implemented for training of NN-Model to calculate process parameters that may end in desired values of the strength factors for given operational condition. The NN-GA codes may be utilized by factory staff to review and assess the association between method parameters and obtained product features, additionally provide operators aided data to enable real time changes.

Rao and Pawar [38] studied factors of optimization of multi-pass milling with goal considered is reduction of processing time exposed to the requirements of arbor quality, arbor redirection and cutting force. The execution of three non-customary evolutionary calculations such as ABC, PSO and SA is considered as far as rate of convergence and exactness of the arrangement. The rate of convergence of ABC and PSO is exceptionally high and these calculations require least iterations for optimal results, though SA requires more iteration.

Lian et al. [39] suggested another methodology called "Similar PSO algorithm (SPSOA)" which depends on PSO and crossover factor to tackle flow shop scheduling problems (FSSP). Despite the fact that optimality is not assured, such a methodology gives arrangements great quality in a sensible time span, and contrasted with GAs, every particles move for best convergence. Performance of the proposed computation is assessed and compare with GA on eight test problems.

Önüt et al. [40] examined an appropriation type distribution center which means to show the issue of structuring a multi-level stockroom considering three dimensional handling cost. One of the commitments of this model is to improve the two-dimensional stockroom plan to the multi-level distribution structure with category based capacity specifically, A, B and C. However, the principle trouble of tackling this sort of planning issues is to battle with nonlinearity in the factors and the imperatives for finding an ideal arrangement. To beat this trouble, they utilized a novel calculation PSO which can discover optimal outcomes in a brief timeframe.

Tandon et al. [41] proposed PSO computation to effectively optimize multiple parameters of milling process simultaneously. Also, this computation is employed for optimization of feed and speed for industrial pocket-milling and minimization of 35% in machining time was measured. GA-based optimization process has been utilized by Palanisamy et al. [42] for machining of mild-steel sample at optimum values of feed, cutting speed associated depth of cut for a continuing MRR in an end-milling process. Similarly Zain et al. [43] implemented GA for the optimization of process parameters of end milling for diminishing surface roughness.

Saravanan et al. [44] observed the optimum machining parameters for continuous profile machining with relevance the minimum cost having practical constraints. They studied power constraint, cutting force and tool-tip temperature as constraints. Because of complex nature of such machining optimization issues SA and GA were implemented to solve.

Bharathi Raja and Baskar [45] delineate empirical models for machining time and surface roughness for optimization of turning process parameters. PSO has been accustomed for optimization of parameters for least machining time under high surface finishing. GA optimization process for resolving multi-pass turning downside is planned by Onwubolu and Kumalo [46]. The outcomes from scrutiny the planned GA-approach with literatures prove its effectiveness. A PSO technique for choosing optimal parameters in multi-pass turning was developed by Srinivas et al. [47] in which PSO is enforced to get the value of cutting parameters that reduces unit cost under practical constraints.

Hecker et al. [48] executed a case study on victimization biological process algorithms to optimize production panning of bakery and modeled production in bakeries as a function of continuous hybrid flow shop. PSO and ACO algorithms were implemented for scheduling issues and to optimize production planning of an example bakery and it was found that both the tools have strength of attain optimized findings for scheduling with in processing time of 15 min.

Jerald et al. [49] proposed optimization method relies on four unconventional computations, i.e., GA, SA, MA and PSO are enforced with success for finding scheduling optimization concerns of FMS. Outcomes are attained for 3 test problems such as ten jobs eight machines, twenty jobs fifteen machines and forty three jobs sixteen machines. Out of all methods PSO outcomes is very promising and provides the best result of objective functions.

Pierreval and Tautou [50] instructed an optimization methodology for production systems. This methodology can be applicable to optimality issues with any variety of variables. This is relies upon combination of biological process rule and a simulation model. Advancements of Michalewicz's biological process operators and rule are projected to handle production system issues. They implemented proposed methodology on test problem: the layout of a workshop manufacturing plastic curd pots. Navalertporn and Afzulpurkar [51] projected hybrid optimization technique utilizing ANN and a bi-directional PSO (BPSO). The projected approach is employed to resolve a process parameter design issues in cement roof-tile production. They concluded that BPSO is an efficient methodology for finding multiple objective optimizations and to solve complex design issues.

Zain et al. [52] projected 2 hybrid systems, integrated SA–GA-type1 and integrated SA–GA-type2, so as to calculate the optimum method parameters of abrasive waterjet machining that result in low machining outcomes. The findings of this investigation showed that hybridization of GA results in calculating optimal process parameters with least machining performance in comparison to real experimental findings.

Yamada et al. [53] projected a layout optimization technique for production cells and an assignment optimization technique for material handling robots in reconfigurable production systems, employing a PSO technique. A GA based computation for resolving facility Layout issues of production system with dynamic features and qualitative and structural call variables was proposed by Azadivar and Wang [54]. The projected approach integrates GAs, computer simulation and an automatic simulation model having friendly interface. Organic process techniques play outstanding role in optimization of supply chain management conjointly. Subramanian et al. [55] projected hybrid forward logistics multi-echelon distribution inventory model (FLMEDIM) and close loop multi-echelon distribution inventory model (CLMEDIM) for the built-to-order surroundings victimization GA and PSO.

Ciurana et al. [56] investigated surface finishing and geometrical and dimensional properties of the grooves/cavities are investigated in laser milling of hardened AISI H13 tool victimization periodic Nd:YAG laser. Additionally PSO is implemented to optimize laser micromachining parameters and this technique will assist in process design conjointly.

Rao et al. [57] investigated single-objective optimisation and multi-objective optimisation factors of an electrochemical machining process parameters employing PSO. This is ascertained that findings of the PSO computation shows important advancement over different optimization processes like goal programing, fuzzy set theory, and GAs. The utilization of GA including the feed-forward NN with back-propagation learning formula for the optimization beneath varied cutting stages of the discharge machining method has been according by Su et al. [58].

Yildiz [59] proposed a new optimization method to resolve optimization issues within the areas of design and production with the integration of PSO and receptor redaction property of immune system and then applied to the optimization of both design concerns and machining parameters taking minimum cost beneath a collection of machining constraints in multi-pass turning operation. An advanced method relies on PSO and local search computation was proposed by Moslehi and Mahnam [60] to resolve the multi-objective flexible job-shop programing drawback with completely different release time. Bean bestowed a strong GA to handle big selection of sequencing and optimization issues like multiple machine programing issues, resource allocation drawback, and quadratic assignment issues [61].

4.3 Genetic Algorithm

This technique was presented by Holland in the year 1975. It is a meta-heuristic pursuit procedure, which works with the idea of Darwin's hypothesis of natural development [62]. GA is a coordinated random pursuit strategy that depends on the mechanics of natural choice and reproducing to effectively investigate a huge space of candidate plans and discover optimal arrangements [63]. GA controls the pursuit via the arrangement region by utilizing natural choice and GA operators like mutation, selection and the crossover.

4.3.1 Principle

GA keeps up a populace of individuals that indicate candidate solutions. Every individual is assessed to give some proportion of its fitness to the issue from the objective function. In every production, another populace is framed by choosing the fittest people dependent on a specific determination system. A few individuals from the new populace experience hereditary tasks to shape new arrangement. The two normally utilized activities are crossover and mutation. After a few productions, the algorithm unites to the best chromosome, which ideally speaks to the optimum or close ideal arrangement. GA has four parts as explained by Davis and Mitchell [64] which are recorded underneath:

1. Mode of encoding obtained values for the issue as chromosome
2. Mode of acquiring an initial populace of arrangements
3. A function that assesses the "fitness" of an answer
4. Reproduction operators for the encoded arrangements.

The well-ordered execution of GA is clarified as follows :

i. **Problem representation**

The first and the principal essential advance in applying GA to an issue is the encoding plan since it can seriously restrain the window of data that has been seen from the framework. To improve the execution of the algorithm, a chromosome portrayal is wanted. By and large, the GA advances a multiple set of chromosomes. The chromosome is generally communicated as a series of factors, every component of which is known as a gene. The factors can be spoken to as real number, binary or different structures and its span is normally characterized by the issue determined.

ii. **Initialization of population**

For initialization of populace, two parameters are used, one is population and other is method to initialize the population. GA cannot depend on single point, rather it produces a number of points having predefined size. Due to which GA has capacity to search from various possibilities of the predefined region and extracts global optima. For normal populace generally size of 20–50 will prefer. Random initiation and heuristic initiation are the two important way of generating initial populace, which randomly produces solution for the complete population.

iii. **Evaluation of fitness function**

The GA imitates the "survival of the fittest" guideline of nature to do the searching and utilizes the fittest value of function as pay off data to direct them via the issue space. When GA knows the ebb and flow proportion of "goodness" about a point, it can utilize this to keep searching ideal. GA is normally reasonable for taking care of maximization issues. Minimization issues are generally changed into maximization issue by some reasonable change.

iv. **Constraint handling**

GA is preferably suitable for unconstrained problems. In any case, the vast majority of the optimization issues are constrained in nature. Henceforth, it is important to change it into an unconstrained issue [65]. Transformation strategies accomplish this by including a penalty term with the objective work. Two primary methodologies for penalty work are: (i) one the basis of violated number of constraints and (ii) in view of some separation from the feasible locale.

v. **Generation of new population**

At that point the assessment ideas are converted into the new populace production to look for the best chromosome in a very natural manner. It comprises of three hereditary factors: (a) Selection, (b) Crossover and (c) Mutation.

(a) **Selection**

This is a process of choosing strings from a populace as per its fitness. The fitness of an individual is assessed concerning a given target function. The most astounding position chromosome will have greater probability of choice and the most exceedingly awful will be dispensed with. There are number of determination strategies accessible. The techniques incorporate, roulette wheel determination, competition choice, position choice, consistent state choice, etc. All in all, "Roulette wheel" determination strategy is utilized. In this technique, parents are chosen by their wellness. The better the chromosomes they have, the more shots are there to be chosen.

(b) **Crossover**

When the determination procedure is finished, next we have to apply crossover operator. Crossover is defined as an operator of recombination which joins subgroups of two parental chromosomes to generate offspring that consists of few sections of both the parental hereditary material. In the hybridization very fit people are offered chances to repeat by trading bits of their hereditary data with other exceptionally fit people. This generates new "offspring" arrangements, which share some great attributes taken from the two guardians. Figure 4 demonstrates the hybridization task between the two parent strings and the formation of off springs. The hybridization factor essentially consolidates substructures of two parent chromosomes to deliver new structures with the picked hybridization probability 'Pc'. It demonstrates how

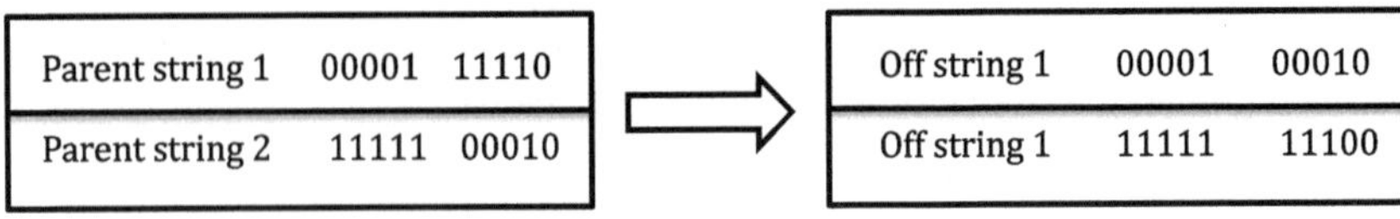

Fig. 4 Crossover operation [66]

Fig. 5 Bitwise mutation [66]

regularly hybridization is performed. A likelihood of 0% implies that the 'offspring' will be the correct imitation of their 'folks' and a likelihood of 100% implies that every production is made out of altogether new spring.

(c) **Mutation**

The selection and crossover administrators will create a lot of various off springs. Be that as it may, there are two primary issues with this. They are

i. Depending upon the initial populace picked, there may not be sufficient decent variety in the underlying strings to guarantee that the GA looks through the whole issue space and
ii. The GA will converge on sub-optima strings because of an awful decision of initial populace.

These issues might be overwhelmed by the presentation of mutation administrator into GA. It is utilized to infuse new hereditary material into the hereditary populace. Transformation can be acknowledged as an arbitrary deformation of the strings with certain likelihood. The beneficial outcome is conservation of hereditary variety and, as an impact that nearby maxima can be maintained a strategic distance from. In this, the offspring can either supplant the entire populace or supplant less fit people. Operator changes 1 as 0 and the other way around by bit wise. Bitwise change is done a little bit at a time by flipping a coin with low likelihood. On the off chance that the result is valid, the bit is changed; generally the bit is not changed.

Greater mutation rate would decimate the fit strings and savage the GA into an arbitrary search. Probability of mutation 'Pm' of 0.01–0.001 is normal and these qualities speak to the likelihood that a specific string will be chosen for change, i.e., for a likelihood of 0.01, one string in one thousand, will be chosen for transformation. Figure 5 delineates the bitwise task. As appeared in Fig. 5 bitwise transformation activity arbitrarily chooses a string and switches the haphazardly picked bit from 0 to 1 or 1 to 0.

vi. **Termination criteria**

Amid the run of algorithm, fitness values increment step by step and at one specific production, fitness value will not increase further which speaks to the optima or close optima arrangement. At this point, running of GA should be stop.

4.3.2 Advantages

Contrasted with conventional continuous optimization strategies, GA has the accompanying critical contrasts.

- GA controls coded forms of the issue parameters rather than the parameters themselves.
- While every ordinary technique seeks from a single point, GA dependably works on an entire populace of focuses (strings). This contributes a lot to the vigor of hereditary calculation. It enhances the shot of achieving the global optima and, the other way around, decreases the danger of getting to be caught in a neighborhood stationary point.
- Normal genetic computations do not utilize any auxiliary data about the target work values like derivatives. Henceforth, they can be connected to any sort of continuous or discrete optimization issue.
- GA utilizes expectation factors while customary techniques for continuous problem apply deterministic factors. All the more explicitly, the manner in which another production is figured from the real one has some arbitrary parts.

4.4 Particle Swarm Optimization

This evolutionary technique was evolved by the combined effort of two scientist of different specialization one is Russell Eberhart and other is James Kennedy in 1995. Kennedy was a social psychologist but Eberhart was an electrical engineer [28]. Kennedy visualizes the intelligence of birds flocking and fish schooling and it was found that they have some intelligence which will give motivation to evolve different evolutionary computation which is called as PSO. This is a characteristics perception that birds can travel in expansive gatherings without impact and they maintain optimum distance between themselves. This segment exhibits a few insights regarding birds in nature and outlines their abilities and sociological conduct also [67].

4.4.1 Principle

The PSO is works on evolutionary computation procedure impersonating the conduct of flocks of birds and their methods for data trade. In PSO various particles are traveled in problem space by an efficient methodology. At time t, every particle i has a vector position, $x_i(t)$, and a vector speed, $v_i(t)$. Memory of PSO stores particle's present location and their personal ever best position. Speed of each particle will vary as per the authentic data put away in the memory and furthermore arbitrary data. Now the new speed will utilized as to update the location of the particle and assess the new position of target function validation.

4.4.2 Initial Solutions

The developed algorithm requires numerous initial points for initiation of search in problem space. These underlying solutions are essentially the particles utilized amid pursuit. Since no particle is conceived or decimated amid the search, the quantity of initial points is actually equivalent to the quantity of particles of the calculation amid its investigation. The developed calculation creates L random numbers as the underlying arrangements which are alluded as x_i; $i = 1, 2, \ldots -L$; where n represents quantity of potential tools in the taken issue. Better diversified particles will be guaranteed by the random permutations of initial points. And producing random numbers in the taken time duration said that feasible search is takes place.

4.4.3 Parameters and Criterion

Developed algorithm considers L particles to investigate the possible space, L particle speeds are additionally expected to refresh position of the each particles amid trails of the calculation. The computation at first creates L arbitrary integers as speed of particle so as to refresh position of the molecule. Consider that speeds must be in a proper span with the goal that the particles stay in feasible region subsequent to being refreshed. Since the feasible arrangement interim is [0; n! − 1], the suitable speed interim which ensures attainability of every particle after refresh in kth cycle for every particle i. Calculation should likewise update particle speeds amid the pursuit to manage the particles with the help of more alluring territories of feasible area. Refereeing to Fig. 6; initially PSO calculation uses to refresh speeds condition as per relation mentioned in equation (ii) [28]:

$$V_k[t+1] = V_k[t] + C_1 r_1 (P_{kbest} - P_k) + C_2 r_2 (G_{kbest} - P_k) \quad (2)$$

where, $V_k[t+1]$ means new velocity of particle,

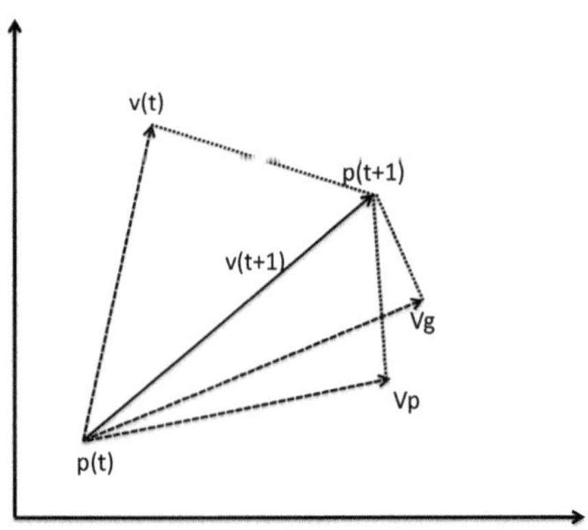

Fig. 6 Vector representation of particle's position and velocity [68]

$V_k[\cdot]$ means old velocity of particle,
P_k means current solution,
P_{kbest} means personal best solution,
G_{kbest} means global best solution,
V_p = velocity of personal best solution,
V_g = velocity of global best solution,
C_1 and C_2 are social and cognitive factors.
Normally, $C_1 = C_2$ in the range of [0–4].
r_1 and r_2 means random numbers between [0–1].

Particle's updated location can be calculated by expression derived in equation (iii):

$$P_k[t+1] = P_k + V_k[t+1] \tag{3}$$

These two equations entail about the new design in search along global optima by applying velocity vector which generates on the basis of local and global optimal point. Hence it is concluded that PSO updates its parameters by learning from previous and neighbors.

4.4.4 Flow Chart

Figure 7 illustrates the flow of the commands for performing particle swarm optimization in MATLAB.

4.4.5 Advantages

PSO is a populace dependent evolutionary method which has many prime favorable circumstances over different strategies as pursues:

- It is a non-derivative algorithm not at all like numerous customary procedures.
- It has the adaptability of joining with other improvement procedures to shape cross breed devices.
- It has few parameters to alter not at all like numerous other contending procedures.
- It can escape local minima.
- Implementation of PSO is very easy and it can program with fundamental numerical and rationale tasks.
- PSO can deal with stochastic target functions as on account of speaking to one of the factors as random.
- PSO do not relies on a selection of good initial value for begin its iterative procedure.

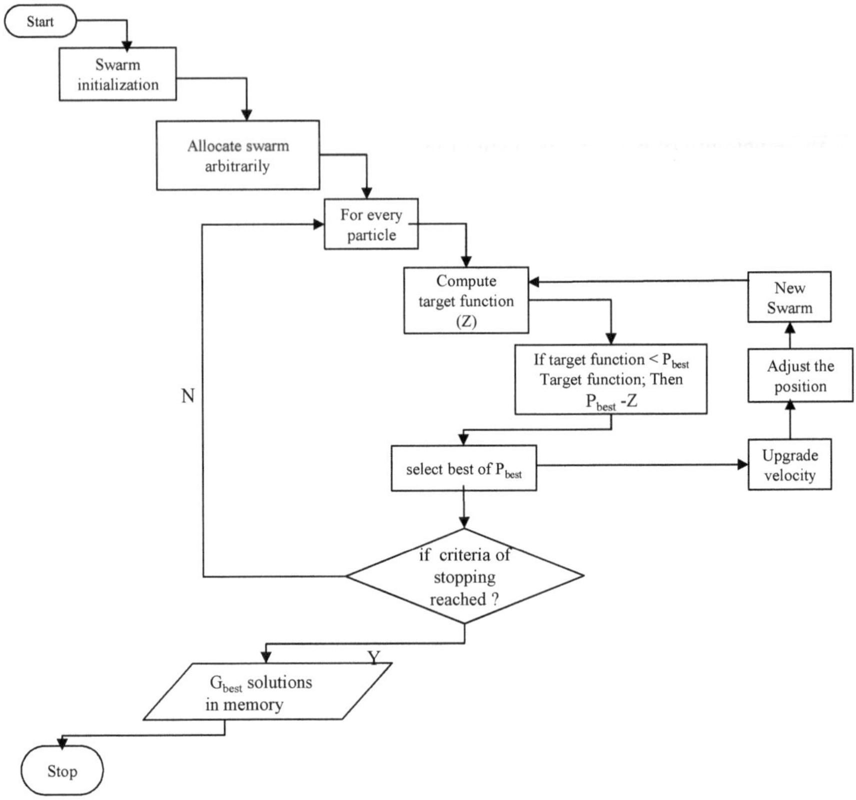

Fig. 7 Flowchart of PSO [69]

5 Case Study: Design of Loop Layout in FMS Using Evolutionary Technique

Loop layout is a common layout in FMS, because it allows placing machines in a loop and transportation of material is along one direction only. A preliminary step in solving loop layout design problem (LLDP) is calculation of machine sequence inside the loop. A LLDP can formulate as permutations of workstations with (m_1, $m_2 \ldots m_n$) considering loading/unloading station with notation 0. Every object is featured by its part route, in sequencing of workstation it must reach to finish processing of job. For a given job, assume processing on jth machine successively processed on ith machine. If location of jth machine is lower than ith, then the job have to traverse through loading/unloading point called as reload. The amount of reloads needed to finish process for a job is outlined as traffic congestion [70]. Afentakis [71] prompt the employment of traffic congestion as deciding factor of the loop layout. The congestion is explained in terms of number of times traverse of job in a loop till complete processing. Mostly two varieties of congestion amount used in LLDP are

MIN_SUM and MIN_MAX [72]. A MIN_SUM drawback makes an attempt to attenuate the whole congestion of all components whereas a MIN_MAX drawback makes an attempt to attenuate the most congestion among family of components.

The effectiveness of the projected algorithm is assessed by the subsequent performance condition mentioned in Nearchou [72].

1. Minimization of average cost of best loop layouts

$$Cost(S) = \sum_{i=1}^{N} reload_i$$

where, S is the best loop layout combination
reload is the crossing through loading/unloading station
N is number of parts.

2. Minimization of average percentage solution effort (%SE) spent by algorithm

$$SE\ (\%) = \left(\frac{NE_{best}}{NE_{total}}\right) \times 100$$

where, NE_{best} is the number of evaluation to get the best result
NE_{total} is the total number of evaluations.

This is required to furnish few numerical coefficients to assess the parameters in PSO and GA because these coefficients directly affect the effectiveness of global optimization. The basic setting of parameters for PSO is given as Size of population = 100, velocity factors = $C_1 = C_2 = 2$, termination criteria = 300 iterations and no. of evaluations = 30,000 {referred from [73]}.

The basic setting of parameters for genetic algorithms is given as P_c (ratio of crossover operation) = 0.3, P_m (ration of mutation operation) = 0.3, pop_size (population size) = 20 and max_gen (maximum generation) = 200 {referred from [70]}.

The PSO computation is evaluated by the randomly crated benchmark issues presented in Nearchou [72]. Required sequencing of workstations of hypothetical test problems are given in Table 1. Findings of PSO computation [69] compared with GA [70, 74] and DEA [72] for the benchmark downside are shown in Table 2. It is observed that for the test problems 1 and 2, PSO results optimal solution as same as other heuristics. In terms of %SE spent by the PSO algorithm to reach the optimal solutions is comparatively lower than DEA_2 and equal to GA for problem, but higher in case of problem 2 due to less number of evaluations. The low amount of %SE indicates the quick convergence of the algorithm towards the optimal solution. The PSO algorithm is executed on MATLAB and graphical representation of the results obtained by Rai and Jayswal [69] are depicted in Figs. 8, 9, 10 and 11. Since computed values in Table 2, is based on the evaluation of layout on traffic congestion factor, but in actual layout every workstation has its own clearance which differs from station to station. Required clearance must be included between each workstation for

Table 1 Required machine sequence [72]

Problem no.	NoM and NoP	Part no.	Required machine sequence
1	10 and 3	1	2–1–6–5–8–9–3–4
		2	10–8–7–5–9–6–1
		3	9–2–7–4
3	15 and 9	1	4–2–5–1–6–8–14–9–11–3–15–12
		2	3–2–15–14–11–1–7–10–4–5–13–6–9
		3	5–6–11–15–2–12–3–4
		4	10–9–4–14–2–3–15–8
		5	11–2–4–14–5–3–15
		6	8–10–12–11–15–13–1–14–4–5–3
		7	5–11–10–3–7–13–8
		8	7–3–2–8–4–10–6–15–13–9–1
		9	11–13–3–1–12–14–4–8–9–2

NoM and NoP—Number of machines and number of parts

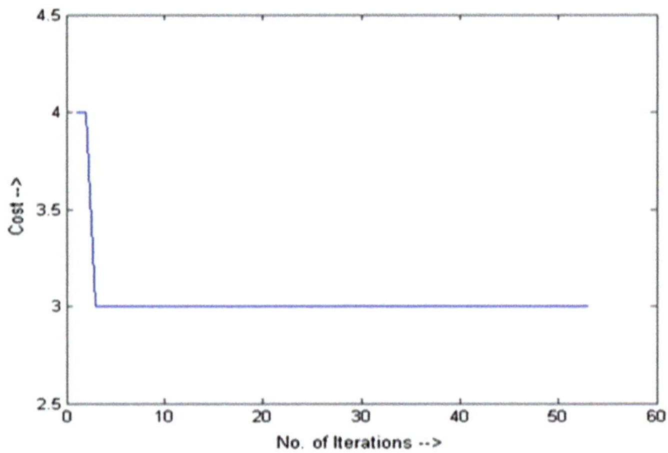

Fig. 8 Graph of number of iterations versus best cost for test problem 1

storage, handling, maintenance and safety. Therefore clearance between workstations must be taken into consideration while solving LLDP [73].

In this case study, an implementation of PSO based approach on obtaining the optimal solution of unidirectional loop layout design problem is discussed. In this work the minimization of total traffic congestion, minimization of total cost in terms of reload and minimization of solution effort have been considered as an objective. The proposed algorithm is tested on different combinations of machines to validate the performance of algorithm, and the obtained results are very promising. It is seen that the PSO algorithm is efficient in finding good quality solutions for the

Table 2 Comparative results of evolutionary techniques for the benchmark problems

Problem no.	NoM and NoP	Algorithm	Cost	SE (%)	Total evaluation	Congestion for each part	Optimal order of machines
1	10 and 3	DEA_2	3	22.5	100	1–2–0	6–5–10–8–9–3–2–7–1–4
		GA	3	1.0	100	1–2–0	10–5–8–9–3–2–7–1–4
		PSO	3	1.0	100	1–2–0	10–8–9–3–2–7–4–1–6–5
2	15 and 9	DEA_2	24	39.5	100	3–4–3–2–2–4–1–4–1	5–7–11–13–10–3–1–6–15–12–8–14–9–2–4
		GA	24	31.6	100	2–4–3–2–2–4–1–4–2	4–7–5–11–13–10–3–1–6–15–8–14–9–2–12
		PSO	24	54	50	2–4–3–3–2–3–1–3–3	7–4–5–11–10–3–15–13–2–1–6–8–12–14–9

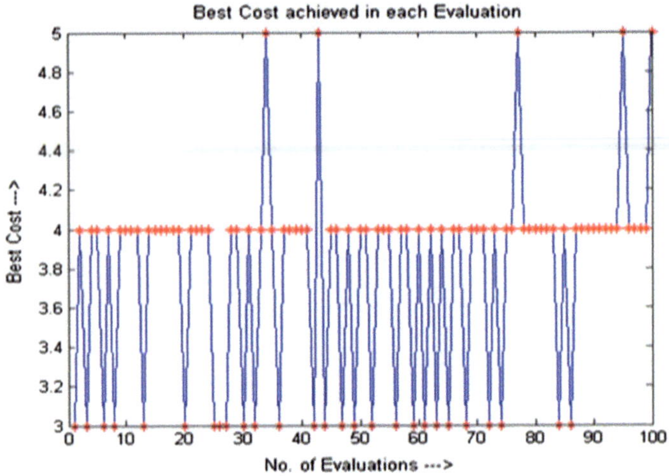

Fig. 9 Graph of total evaluation versus best cost for test problem 1

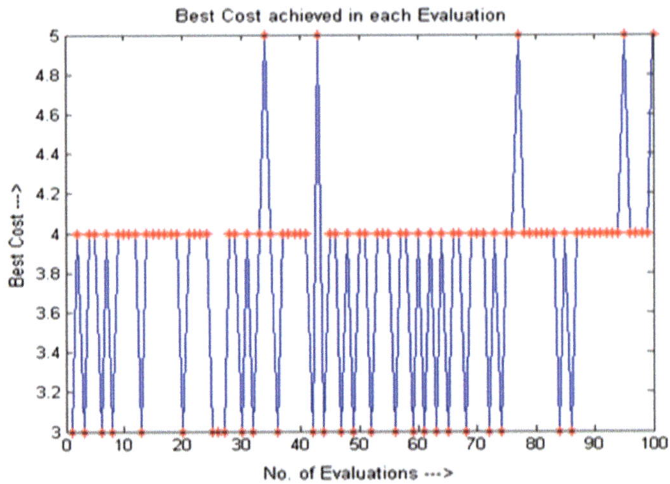

Fig. 10 Graph of number of iterations versus best cost for test problem 2

layout problems with less percentage solution effort. A comparison chart with other techniques such as GA and DEA also suggested that evolutionary approach in manufacturing leads towards the optimality in each and every aspect of the manufacturing technology. So the present era of manufacturing needs implementation of various evolutionary techniques to resolve these issues.

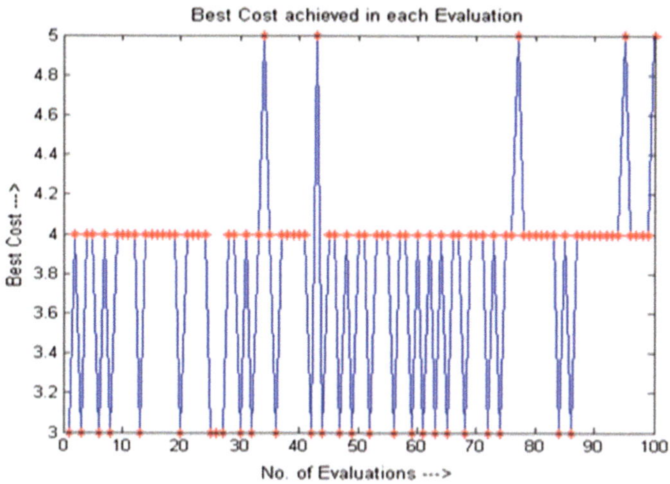

Fig. 11 Graph of total evaluation versus best cost for test problem 2

6 Comparisons Between PSO and GA

There are a few similitudes among GA and PSO. Both these calculations begin with a populace of arrangements created randomly and the nature of these arrangements is communicated as far as their fittest values. There are a few dissimilarities additionally among PSO and GA. For instance, in PSO, there are no crossover and mutation parameters, though these are considered as critical factors of the GA. In PSO, the particles have memory, and thusly, effectively discovered great data of the particles is conveyed forward iteratively. Then again, the past learning of the issue is lost once the populace changes. A GA is an amazing asset for global optimization. Then again, PSO completes both the local and global seek at the same time. PSO calculation is more straightforward in development and quicker than the GA. The PSO may give more precise outcomes contrasted with the GA.

7 Future Aspects of Evolutionary Techniques

Evolutionary calculations are nature motivated populace based optimization strategies, yet they have a few constraints in either perspective. Because of this reality, extensive research is needed to verify computations for various issues to assess their reasonableness for a large assortment of issues. Research is kept on upgrading the current computations to enhance their execution. Improvement can happen either (a) by changing the current computation methods or (b) by hybridization of the current computation methods. Improvement because of alterations in the current computation is accounted for in GA [75], PSO [76], ACO [77], ABC [78], etc. Upgrade should likewise be possible by joining the qualities of various optimization computations, called as hybridization of computations. It is a compelling method to make the computation proficient and it consolidates the features of various computations.

References

1. Nocedal J, Wright SJ (2006) Numerical optimization, 2nd edn, Springer
2. Deb K, Sindhya K, Hakanen J (2016) Multi-objective optimization. In: Decision sciences: theory and practice
3. Karpinski A, Wink R (2012) Manufacturing processes. In: Industrial high pressure applications: processes, equipment and safety
4. Kalpakjian S, Schmid SR, Musa H (2009) Manufacturing engineering and technology
5. Groover MP (2002) Automation, production systems and computer-integrated manufacturing
6. Mehrabi MG, Ulsoy AG, Koren Y (2000) Reconfigurable manufacturing systems: key to future manufacturing. J Intell Manuf
7. Molina A et al (2005) Next-generation manufacturing systems: key research issues in developing and integrating reconfigurable and intelligent machines. Int J Comput Integr Manuf
8. Womack J, Jones D, Roos D (1990) The machine that changed the world : the story of lean production
9. Adam EE (1983) Towards a typology of production and operations management systems. Acad Manag Rev
10. Up B et al (2006) Cellular manufacturing. Manuf Eng
11. Fulkerson B (1997) A response to dynamic change in the market place. Decis Support Syst
12. Warnecke HJ, Steinhilper R (1982) Flexible manufacturing systems; new concepts; EDP-supported planning; application examples
13. Koren Y et al (1999) Reconfigurable manufacturing systems. CIRP Ann Manuf Technol
14. Tolio T (2009) Design of flexible production systems: methodologies and tools
15. Selladurai RS (2004) Mass customization in operations management: oxymoron or reality? Omega 32(4):295–300
16. Hon KKB (2005) Performance and evaluation of manufacturing systems. CIRP Ann Manuf Technol
17. Lanza G, Peters S, Herrmann H (2012) Dynamic optimization of manufacturing systems in automotive industries. CIRP J Manuf Sci Technol 5(4):235–240
18. Huang SH et al (2002) Manufacturing system modeling for productivity improvement. J Manuf Syst
19. Jayal AD, Badurdeen F, Dillon OW, Jawahir IS (2010) Sustainable manufacturing: modeling and optimization challenges at the product, process and system levels. CIRP J Manuf Sci Technol
20. Kelley TR (2010) Optimization, an important stage of engineering design. Technol Teach 69(5):18–23
21. Merrill C, Custer RL, Daugherty J, Westrick M, Zeng Y (2008) Delivering core engineering concepts to secondary level students. J Technol Educ
22. Taha HA (2005) Operations research: an introduction
23. Yoshimura M (2010) System design optimization for product manufacturing
24. Jahangirian M, Eldabi T, Naseer A, Stergioulas LK, Young T (2010) Simulation in manufacturing and business: a review. Eur J Oper Res
25. Kolda TG, Lewis RM, Torczon V (2003) Optimization by direct search: new perspectives on some classical and modern methods. SIAM Rev
26. Krishnamoorthy K (1991) Estimation of normal covariance and precision matrices with incomplete data. Commun Stat Theory Methods

27. Rao RV, Savsani VJ (2012) Mechanical design optimization using advanced optimization techniques
28. Kennedy J et al (1995) Particle swarm optimization. In: Proceedings of the IEEE international conference on neural networks, 1995
29. Mirjalili S (2019) Genetic algorithm. In: Studies in computational intelligence
30. Lampinen JA, Price KV, Storn RM (2005) Differential evolution: a practical approach to global optimization
31. Dorigo M, Di Caro G (1999) Ant colony optimization: a new meta-heuristic. In: Proceedings of the 1999 congress on evolutionary computation, CEC 1999
32. Mobini M, Mobini Z, Rabbani M (2011) An artificial immune algorithm for the project scheduling problem under resource constraints. Appl Soft Comput J
33. Karaboga D, Basturk B (2007) A powerful and efficient algorithm for numerical function optimization: artificial bee colony (ABC) algorithm. J Glob Optim
34. Blum C et al (2012) Evolutionary optimization. In: Variants of evolutionary algorithms for real-world applications
35. Bulgak AA, Diwan PD, Inozu B (1995) Buffer size optimization in asynchronous assembly systems using genetic algorithms. Comput Ind Eng 28(2):309–322
36. Ozcelik B, Erzurumlu T (2006) Comparison of the warpage optimization in the plastic injection molding using ANOVA, neural network model and genetic algorithm. J Mater Process Technol 171(3):437–445
37. Cook DF, Ragsdale CT, Major RL (2000) Combining a neural network with a genetic algorithm for process parameter optimization. Eng Appl Artif Intell 13(4):391–396
38. Rao RV, Pawar PJ (2010) Parameter optimization of a multi-pass milling process using non-traditional optimization algorithms. Appl Soft Comput J 10(2):445–456
39. Lian Z, Gu X, Jiao B (2008) A novel particle swarm optimization algorithm for permutation flow-shop scheduling to minimize makespan. Chaos, Solitons Fractals 35(5):851–861
40. Önüt S, Tuzkaya UR, Doğaç B (2008) A particle swarm optimization algorithm for the multiple-level warehouse layout design problem. Comput Ind Eng 54(4):783–799
41. Tandon V, El-Mounayri H, Kishawy H (2002) NC end milling optimization using evolutionary computation. Int J Mach Tools Manuf 42(5):595–605
42. Palanisamy P, Rajendran I, Shanmugasundaram S (2007) Optimization of machining parameters using genetic algorithm and experimental validation for end-milling operations. Int J Adv Manuf Technol 32(7–8):644–655
43. Zain AM, Haron H, Sharif S (2010) Application of GA to optimize cutting conditions for minimizing surface roughness in end milling machining process. Expert Syst Appl 37(6):4650–4659
44. Saravanan R, Asokan P, Vijayakumar K (2003) Machining parameters optimisation for turning cylindrical stock into a continuous finished profile using genetic algorithm (GA) and simulated annealing (SA). Int J Adv Manuf Technol 21(1):1–9
45. Bharathi Raja S, Baskar N (2011) Particle swarm optimization technique for determining optimal machining parameters of different work piece materials in turning operation. Int J Adv Manuf Technol 54(5–8):445–463
46. Onwubolu GC, Kumalo T (2001) Optimization of multipass turning operations with genetic algorithms. Int J Prod Res 39(16):3727–3745
47. Srinivas J, Giri R, Yang SH (2009) Optimization of multi-pass turning using particle swarm intelligence. Int J Adv Manuf Technol 40(1–2):56–66
48. Hecker FT, Hussein WB, Paquet-Durand O, Hussein MA, Becker T (2013) A case study on using evolutionary algorithms to optimize bakery production planning. Expert Syst Appl 40(17):6837–6847
49. Jerald J, Asokan P, Prabaharan G, Saravanan R (2005) Scheduling optimisation of flexible manufacturing systems using particle swarm optimisation algorithm. Int J Adv Manuf Technol 25(9–10):964–971
50. Pierreval H, Tautou L (2007) Using evolutionary algorithms and simulation for the optimization of manufacturing systems (December 2014):37–41

51. Navalertporn T, Afzulpurkar NV (2011) Optimization of tile manufacturing process using particle swarm optimization. Swarm Evol Comput 1(2):97–109
52. Zain AM, Haron H, Sharif S (2011) Optimization of process parameters in the abrasive waterjet machining using integrated SA-GA. Appl Soft Comput J 11(8):5350–5359
53. Yamada Y, Ookoudo K, Komura Y (2003) Layout optimization of manufacturing cells and allocation optimization of transport robots in reconfigurable manufacturing systems using particle swarm optimization. In: Proceedings 2003 IEEE/RSJ international conference on intelligent robots and systems (IROS 2003) (Cat. No. 03CH37453), vol 2, no October, pp 2049–2054
54. Azadivar F, Wang JJ (2010) Facility layout optimization using simulation and genetic algorithms. Int J Prod Res (March 2013):37–41
55. Subramanian P, Ramkumar N, Narendran TT, Ganesh K (2012) A technical note on analysis of closed loop supply chain using genetic algorithm and particle swarm optimisation. Int J Prod Res 50(2):593–602
56. Ciurana J, Arias G, Ozel T (2009) Neural network modeling and particle swarm optimization (PSO) of process parameters in pulsed laser micromachining of hardened AISI H13 steel. Mater Manuf Process 24(3):358–368
57. Rao RV, Pawar PJ, Shankar R (2008) Multi-objective optimization of electrochemical machining process parameters using a particle swarm optimization algorithm. Proc Inst Mech Eng B J Eng Manuf 222(8):949–958
58. Su JC, Kao JY, Tarng YS (2003) Optimisation of the electrical discharge machining process using a GA-based neural network. Int J Adv Manuf Technol 1(1):1
59. Yildiz AR (2009) A novel particle swarm optimization approach for product design and manufacturing. Int J Adv Manuf Technol 40(5–6):617–628
60. Moslehi G, Mahnam M (2011) A Pareto approach to multi-objective flexible job-shop scheduling problem using particle swarm optimization and local search. Int J Prod Econ 129(1):14–22
61. Bean JC (1994) Genetic algorithms and random keys for sequencing and optimization. ORSA J Comput 6(2):154–160
62. Holland JH (1992) Adaptation in natural and artificial systems. University of Michigan Press, Ann Arbor, MI
63. Goldberg DE, Holland JH (1988) Genetic algorithms and machine learning. In: Machine learning
64. Davis LD, Mitchell M (1991) Handbook of genetic algorithms
65. Deb K (1999) Multi-objective genetic algorithms: problem difficulties and construction of test problems. Evol Comput
66. Phatak AM, Pande SS (2012) Optimum part orientation in rapid prototyping using genetic algorithm. J Manuf Syst
67. Blum C, Groß R (2015) Swarm intelligence in optimization and robotics. In: Springer handbook of computational intelligence
68. Poli R, Kennedy J, Blackwell T (2007) Particle swarm optimization. An overview. Swarm Intell
69. Rai RS, Jayswal S (2018) Design and optimization of loop layout in flexible manufacturing system using particle swarm optimization. Int J Adv Technol 9(2)
70. Cheng R, Gent M, Tosawa T (1996) Genetic algorithms for designing loop layout manufacturing systems. Comput Ind Eng
71. Afentakis P (1989) A loop layout design problem for flexible manufacturing systems. Int J Flex Manuf Syst
72. Nearchou AC (2006) Meta-heuristics from nature for the loop layout design problem. Int J Prod Econ 101(2):312–328
73. Satheesh Kumar RM, Asokan P, Kumanan S (2008) Design of loop layout in flexible manufacturing system using non-traditional optimization technique. Int J Adv Manuf Technol 38(5–6):594–599
74. Cheng R, Gen M (1998) Loop layout design problem in flexible manufacturing systems using genetic algorithms. Comput Ind Eng 34(1):53–61

75. Wongrat W, Srinophakun T, Srinophakun P (2005) Modified genetic algorithm for nonlinear data reconciliation. Comput Chem Eng
76. Petalas YG, Parsopoulos KE, Vrahatis MN (2007) Memetic particle swarm optimization. Ann Oper Res
77. Huang SH, Lin PC (2010) A modified ant colony optimization algorithm for multi-item inventory routing problems with demand uncertainty. Transp Res E Logist Transp Rev
78. Karaboga D, Akay B (2011) A modified artificial bee colony (ABC) algorithm for constrained optimization problems. Appl Soft Comput J

Index

K. Gupta and M. K. Gupta (eds.), *Optimization of Manufacturing Processes*, Springer Series in Advanced Manufacturing, https://doi.org/10.1007/978-3-030-19638-7

MIX
Papier aus verantwortungsvollen Quellen
Paper from responsible sources
FSC® C105338

If you have any concerns about our products, you can contact us on
ProductSafety@springernature.com

In case Publisher is established outside the EU, the EU authorized representative is:
Springer Nature Customer Service Center GmbH
Europaplatz 3, 69115 Heidelberg, Germany

Printed by Libri Plureos GmbH
in Hamburg, Germany